## *About the author*

Fred Roberts has lived in Britain throughout the years covered by this book, much of them spent in the service of the government. After post-graduate studies in fuel technology, he left university in 1943 to work for the Ministry of Fuel and Power, advising on more efficient ways of using fuel and power in industry and commerce in order to help alleviate an acute national coal shortage. 1950 found him at Windscale Works (Sellafield) in Cumbria, working as a development manager in the research and development laboratories of the Ministry of Supply's Department of Atomic Energy where he played an important role in ensuring the success of Operation Hurricane.

The twenty-three years from 1954 to 1977 were spent at the Atomic Energy Research Establishment at Harwell where he obtained successive promotions in research management. He headed research teams studying spent nuclear fuel reprocessing, graphites for nuclear reactor moderator and fuel containment. He was also responsible for certain non-nuclear research projects including carbon fibre development in association with the Royal Aircraft Establishment, Farnborough.

After a sabbatical year in 1973-4 as Visiting Professor at Brunel University with a remit to study materials and energy resources, he returned to Harwell to become deputy head of the Department of Energy's newly-formed Energy Technology Support Unit but continued his involvement with the university as a Professor Associate. From 1977 until retirement in 1986, he was an independent consultant often working on energy analysis projects for the Department of Energy and also the private sector of industry.

The author sat on many government committees concerned with energy and was the UK delegate to the NATO committee for challenges in modern society when it carried out an examination of energy consumption and use by industrial sectors. He has had many technical reports and reviews published and has read papers on physical resources, materials technology and energy at various scientific institutions and universities. He is a Fellow of the Institution of Chemical Engineers.

# Sixty Years of Nuclear History

## Britain's Hidden Agenda

FRED ROBERTS

JON CARPENTER

Our books may be ordered from bookshops or (post free) from
Jon Carpenter Publishing, 2 The Spendlove Centre, Charlbury,
England OX7 3PQ

Please send for our free catalogue

Credit card orders should be phoned or faxed to 01689 870437
or 01608 811969

Our US distributor is Paul and Company, PO Box 442, Concord, MA 01742
(to order, phone 201 840 4748, fax 201 840 7148)

First published in 1999 by
Jon Carpenter Publishing
2 The Spendlove Centre, Charlbury, Oxfordshire OX7 3PQ
☎ 01608 811969

ISBN  1 897766 48 3

Printed in England by J. W. Arrowsmith Ltd., Bristol.
Cover printed by KMS Litho, Hook Norton

# Contents

## Diagrams

# Acknowledgements

My special thanks are due to Tony Roberts for many hours of editing, preparing the two illustrations and also for compiling the index. Thanks also to Joyce Roberts for many very valuable suggestions on the handling of material, and to Chris Roberts for supplying material for the book.

I am grateful to Professor Paul Rogers of Bradford University for supplying information about nuclear weapons and to Sheila Jones for the loan of material from CND archives.

Tony Benn is to be thanked for commenting on the parts of the book which refer to the times when he was a member of the cabinet as Energy Secretary.

Peter Tatton-Brown (Capt RN retd) has kindly read and commented on chapters concerned with nuclear weapons and submarines and John Roberts (Capt RN retd) has also commented on parts of the book.

Thanks also to Bill Arnold, a physicist, for editing the Glossary.

Dr Frank Stainthorp of Architects and Engineers for Social Responsibility provided valuable help in supplying me with scientific information and also in giving me assistance by reading the manuscript and supplying detailed criticism.

Finally, my thanks go to those who read the script at the half way stage in 1995 and encouraged me to go on and complete the task.

F. R.

# Preface

Prior to the discovery of the microchip, nuclear power was usually regarded as being the major discovery of the century. Billions of pounds of taxpayers' money have been poured into exploiting this energy source for military weapons and electricity generation alike. Much has been kept from the public over what has been done with vast amounts of their money in their name, ostensibly for reasons of security. Was the money well spent? Did nuclear energy developments denude other projects of money, skilled manpower or physical resources? Were the problems and dangers of radioactivity underestimated, sometimes ignored? If, as we go into a new century, oil and gas become scarce and, despite the availability of renewable energy sources such as wind and solar, the country is tempted to launch a second nuclear programme, it will be vital to avoid the pitfalls of the first one.

Whilst working in the nuclear business it dawned upon me how closely linked were the two strands of nuclear power — electricity generation and nuclear weapons production. I became aware of how the links had both a technical and a political dimension. Many books have been written on either nuclear electricity or nuclear weapons, generally covering a fixed period of time. But now I consider it imperative for there to be available a concise, analytical history of *both* strands of nuclear power in Britain, starting at the beginning of World War II and continuing to the present time, a span of about sixty years. In 1939, this country was still in the *pre-nuclear age* but in the late nineties we are entering the *post-nuclear age*. J. S. Mill once said that politics concerns those subjects on which it is in the interests of rulers that people should be misled. In my view, this definition certainly applies to the development of nuclear energy and this book is therefore concerned more with the *political* history of the exploitation of nuclear fission than with its technology. It should be stressed that I am not using the word politics in the sense of parliamentary politics. Indeed, over the years there has been very little debate in Britain's parliament concerning nuclear affairs because of the ongoing cross-bench agreement on such matters, irrespective of whether there was a Labour or Conservative government in power. This agreement can be traced right through from Winston Churchill's war-time coalition to Tony Blair's administration. At the outbreak of World War II, nuclear fission had just been discovered along with its potential for releasing an enormous amount of energy from the atom and it was perhaps inevitable that embattled Britain would soon consider the possibility of developing a bomb with devastating power. It gradually became clear that she did not have the necessary resources and the large size of production plants would make them vulnerable to aerial attack by Germany; so Churchill turned to the US government for help and also revealed to them all our secret nuclear information. The Americans soon began to develop a nuclear bomb project of their own which went ahead with great speed when they entered the war after the Japanese bombing of Pearl Harbour. But the project included a mission of nuclear scientists from Britain, many of the senior scientists being Jewish refugees from mainland Europe under the Nazis. The eventual outcome of this highly secret project became known to the world when two nuclear bombs destroyed the Japanese cities of Hiroshima and Nagasaki. Almost immediately after the war, Clement Attlee and certain of his ministers began to plan a secret nuclear bomb programme in Britain. Now this seems a strange thing to have done, for the war had exhausted her and she was drained of economic and industrial resources to a high degree. The Americans were appealed to for financial aid and Britain obtained a $3.75 billion loan, although this was only granted on very onerous terms. In establishing her own programme Britain needed all the information she could get on nuclear technology and expected to get this from the US. But it was not to be so, because the McMahon Act was passed and this prevented the flow of nuclear information outside the US.

In 1948, the Berlin airlift resulted in Britain giving the US permission to station strategic nuclear bombers over here. This was a dangerous action to take because just previous to this we had agreed a modus vivendi with the Americans which would allow us to obtain more nuclear information from them but at the same time Britain surrendered its right to veto American use of the nuclear bomb. Thus at the start of the Cold War there were American bombers on our soil, maybe with nuclear weapons ready to take off for Russia if the Pentagon so decided. But it is quite possible that the Russians would soon have known of the presence of the US nuclear bombers so close to them. The Russians had their own nuclear project and this was so advanced that they were able to explode a nuclear device as early as 1949; Britain did not do so until 1952.

A study of Britain's nuclear history reveals how often Anglo-American relations crop up. In the middle 1950s, Britain sent a team to the US to learn the technology of building nuclear submarines. In 1962, Britain realised that she lacked the resources to go on developing rockets and therefore chose to become dependent on the US to provide her with the means of delivering strategic nuclear and thermonuclear warheads. Hence we bought the Polaris weapon and following that, Trident.

Throughout the years, many of the facts concerning Britain's nuclear weapons programme have been hidden from the public and only some have seen the light of day many years later. This even applied to the so-called peaceful use of nuclear energy, the nuclear power stations programme and the reprocessing of its spent fuel elements to extract plutonium. Some of this plutonium became involved in a secret deal and was shipped across the Atlantic to the USA. Whether or not this material was actually used in weapon production over there, it nevertheless resulted in American nuclear weapon materials being sent to Britain in exchange.

The source of any nuclear programme, whether for bomb production or civil electricity supply, lies in having available a supply of naturally occurring uranium ore. There was the strange story relating to a government contract with RTZ for uranium ore supplies — were these not coming to Britain from Namibia in conflict with a United Nations ruling?

After World War II, once it became general knowledge that Britain was manufacturing nuclear warheads, there was a growth in awareness of the health and safety problems associated with the nuclear industry and also the beginnings of a long-standing debate on the moral aspects of the bomb. Concern over aerial testing of bombs by the US, UK and the Soviet Union in the 1950s eventually led to public protests and in Britain to the formation of the Campaign for Nuclear Disarmament (CND). At the international level, early discussions at the United Nations about nuclear disarmament had failed in 1948. Later attempts were frustrated by various international crises — Suez and Hungary, both in 1956, then the Berlin Crisis in 1961 and finally the Cuban Missile incident in 1962. This book describes the gradual rise to pre-eminence of an over-confident, subsidised nuclear electricity industry, only to see it fall when the true costs and risks emerged and the City passed judgement on it. It is seventeen years since an application was filed for the building of Sizewell B, the last nuclear power station to be built in Britain. However, civil and military nuclear power stations still continue to operate in this country and so long as they do, and their spent fuel elements are processed in the Sellafield chemical plants, so will the quantities of dangerous radioactive wastes continue to increase. Furthermore, spent fuel elements from abroad are also processed at Sellafield. This has given rise to the media referring to Britain as The Dustbin of Europe. Unfortunately, the technology for dealing safely with high level radioactive wastes on a long term basis is not yet established. It therefore looks as though future generations will be left to grapple with this legacy from our venture into using nuclear power for peaceful and military purposes in this century. The ending of the Cold War in the early 1990s signalled a decline in demand for nuclear weapons by the NATO and Warsaw Pact powers but there is the serious problem of dealing with the nuclear materials contained in the remaining nuclear and thermonuclear warheads. Also, there are other countries which have nuclear weapons of unknown quantity.

Each chapter of the book treats a particular subject and sequence of events as a story in itself, thus there is some unavoidable overlapping in time scale. The first chapter is of an introductory nature for it gives a short account of the history of the discovery of radioactivity and the development of nuclear science, prior to 1939 and the outbreak of World War II. Some key scientific terms inevitably arise in a book of this nature but these are kept to the minimum and in any case they are defined in a short glossary. To make for easy reading there are no references in the text but a short bibliography is included. Finally, it is hoped that this book will provide vital reading to the lay reader as well as students of the history and politics of the nuclear era in Britain. It should be of interest to American and Commonwealth readers. It spans six disturbing decades of history.

Part I

# BRITAIN AND NUCLEAR WEAPONS: THE EARLY YEARS

# Chapter 1

# Atomic science to 1939

It was not until 1939 that sufficient knowledge about the atom had been acquired to suggest the possibility of putting it to practical use, for good or evil; as a major war was then starting it was perhaps inevitable that an early attempt would be made to apply it for warlike purposes. Nuclear politics, the subject of this book, had been born; the genie was out of the bottle and the development and manufacture of two nuclear bombs was relentlessly pursued, culminating in the destruction in August 1945 of two Japanese cities. However, before starting to explore the politics of the atom from World War II and on through succeeding decades, it may be interesting and useful to review briefly the history of atomic science leading up to the situation in 1939. But those with little interest in this subject may move straight on to Chapter 2. A glossary of some key technical terms is appended to the book for reference purposes.

The idea that matter is composed of tiny, discrete particles was probably put forward initially by early Greek philosophers. There were two opposing views around in the 5th century BCE. Some such as Democritus and Epicurus taught that matter was composed of very tiny particles, atoms, which could not be further subdivided. But Aristotle preached that matter was continuous and that all substances could be formed from four basic elements — earth, air, water and fire. Both theories were based on very slender experimental evidence and so neither could be taken very seriously.

However, in medieval times, the alchemists appeared to adopt Aristotle's theory and believed that by changing the proportion of these elements in any one substance they could transmute it into another one; to them, the possibility of transmuting lead into gold was the 'Holy Grail'.

## Origin of the Periodic Table

In the sixteenth century, the alchemists' list of elements was increased by Paracelsus, a Swiss physician, who added mercury, sulphur and salt, but removed fire. In the next century, Robert Boyle, a British philosopher and chemist, postulated that many more substances such as gold, silver and lead together with 'airs' such as hydrogen, oxygen and nitrogen should be regarded as elements. He even added a new one to the list — phlogiston. But this led to much confusion, eventually sorted out by a French chemist, Antoine Lavoisier, in the eighteenth century, who declared phlogiston redundant. Lavoisier compiled a list of some 33 elements, most of which are still recognised as such. He has been said to have laid the foundations of modern chemistry by giving a correct explanation of the part played by oxygen in the process of combustion.

At the beginning of the nineteenth century, John Dalton, a British chemist, arrived at the basic conclusion that elements were in fact atoms and capable of combining with each other in simple proportion. He related these proportions to one another, using hydrogen as a base reference. Thus did he produce a list of elements in increasing order of mass. Dalton was awarded a medal of the Royal Society in 1826 in recognition of what became called 'the atomic theory of chemistry'. Other chemists continued to add significantly to the list of elements, determined their atomic weights and indeed by 1860 a total of 70 diverse elements were known. In 1869 the Russian chemist Dimitri Mendeleev published a definitive classification known as the Periodic Table. A modern textbook on inorganic chemistry would provide the present reader with an extensive review of the features of the Periodic Table where particular mention is made that atomic masses relate to oxygen as 16 and the atomic numbers are the numbers of the elements in the sequence. In Mendeleev's time, the table stopped at the 85th element, bismuth, but in the last decade of the nineteenth century a series of elements with atomic masses greater than 209 were found by A. H. Becquerel (French), Marie Curie (Polish) and her husband Pierre Curie (French), among others.

Although the atomic masses in the Periodic Table were nearly always, to a first approximation, an integer multiple of the atomic mass of hydrogen, chemists in the early years of this century worried about some notable exceptions. For instance, chlorine had an atomic mass of 35.46. Then the work of F. Soddy and H. Moseley in interpreting the X-ray spectra of anomalous elements showed that it was possible to have atoms of different atomic masses but identical chemical properties. These Soddy referred to as *isotopes* and in the case of chlorine it was deduced that the naturally occurring element was a mixture of two *isotopes* — one with an atomic mass of 35 and the other with 37.

### Discovering the characteristics of radioactivity

In the spring of 1896 Becquerel was trying to elucidate the phenomenon of phosphorescence by exposure of uranium crystals to sunlight. However, a photographic plate which had been wrapped in thick black paper, on developing, showed the clear outlines of a coin which had accidentally lain between the plate and a dish of uranium salts. He concluded that the uranium solution was emitting radiation similar to the so-called 'black light' which had been discovered earlier in the year by Röntgen in Germany and which was called Röntgen rays or X-rays. Marie Curie embarked on a study of this phenomenon and soon found that pitchblende, the mineral source of uranium, was more radioactive than the uranium. She concluded that there must be at least one more element present that was even more radioactive than the uranium. She went on to isolate two new elements, polonium and radium. Her work also showed that two of the naturally occurring radioactive elements, uranium and thorium, disintegrated by a series of stages until the end element, lead, was reached.

Ernest Rutherford, later to become Lord Rutherford, a New Zealand physicist who was working at Manchester in the early years of this century, found that the radiation emitted from uranium was of two kinds which he called 'alpha rays' and 'beta rays'. Then a third kind of radiation known as 'gamma rays' was discovered. Rutherford and his co-worker, Soddy, claimed that radioactive atoms were spontaneously and continuously ejecting part of themselves and thereby changing into a different kind of atom. In a way, it was the transmutation of elements which the alchemists in the past had been searching for in their quest for a way to change base metals into gold! A good deal of this work was made possible by completely novel experimental techniques developed by physicists such as J. J. Thomson, based on passing a current through a gas contained in a glass tube very much like the TV cathode ray tube of today. It was with such equipment that the discovery was made of the existence of particles much smaller than atoms and the physicists began to suspect that the atom may have some kind of structure and was perhaps even penetrable.

By 1910 the general characteristics of radioactivity were known. Perhaps the most important of these is what is termed the 'half-life' of a radioactive element. This is the time taken for half of the radioactive atoms present in a sample to decay by emitting their particles with the other atoms still unchanged and therefore still radioactive. Then, of the radioactive atoms left unchanged, half of these will decay over the same length of time, and so on. Different kinds of radioactive atoms have different half-life periods; these can vary enormously, from a millionth of a second to millions of years. In 1935, Dempster would show that the predominant isotope of uranium has a mass of 238 units but it also contains a small proportion of the lighter isotope of mass 235, which is in fact present at about 0.7 per cent. This single fact was to become of great significance in the military and commercial development of nuclear energy.

## A new model of the atom

In 1911 Rutherford postulated that atoms of all elements had a structure in which most of the mass was contained in a minute region at the centre, which he called the *nucleus* of the atom. This tiny nucleus carried a positive electrical charge and negatively charged electrons moved round it in distant orbit, rather like planets around the sun. The positive charge on the nucleus equalled the total negative charge of the orbiting electrons. Most of the atom was empty space; this was a far cry from the tiny billiard balls envisaged by Dalton and the eighteenth century atomists.

Rutherford's theory did not at first have a great deal of impact on the physicists until Niels Bohr came over to Manchester from Denmark to work beside him for some months in 1912. Bohr's work in theoretical physics greatly firmed up the theory. The model of the atom now postulated gave a perfect and rational explanation for the positions and properties of the elements as found in the Periodic Table. The existence of isotopes could now be explained by each of the nuclei being constructed from positively charged protons together with an equal number of uncharged particles having the same mass (now called *neutrons*) plus a number (usually even) of extra neutrons.

## Disintegrating the nucleus

The chemical properties of each element are determined by the number of electrons in the (outer) orbits whilst the atomic mass is determined almost entirely by the nucleus. The alpha particles discharged from radioactive elements consist simply of two protons combined with two neutrons. Simple electrical theory would lead one to expect the two protons, being positively charged, to repel one another. The fact that they don't is explained by the existence of 'nuclear binding forces' that manifest themselves when the protons and neutrons become in very close proximity to one another.

Rutherford had by now begun to carry out sophisticated experiments in which he bombarded the atomic nucleus with either very fast electrons or alpha particles. Soon he succeeded in disintegrating the nucleus of the nitrogen atom using a stream of alpha particles, thus ejecting from it a hydrogen atom nucleus and leaving behind an isotope of oxygen. This was probably Rutherford's crowning achievement at Manchester and soon he moved to Cambridge where he set up a research school which was said to have dominated nuclear physics for the next twenty years or so.

In the Cavendish Laboratory at Cambridge, Rutherford went on to bombard various types of atomic nuclei. Quite often a hydrogen atom nucleus was knocked out, and this was identified as the proton. Then James Chadwick identified the neutron so that all three of the fundamental particles which made up the atom — proton, neutron and electron — had been positively identified.

The 1920s and early thirties saw further advances in our understanding of matter. We now realise that things are much more complex than we thought. Much is owed to developments in mathematics. In 1927 George Thomson, son of J. J. Thomson who had discovered the electron, proved theoretically that electrons behave both as particles and as *waves*, although it was not until 1987, in Tokyo, that a practical experiment was successfully carried out to support this. However particle-wave duality, quantum mechanics and the latest theories about the structure of matter are well outside the scope of this book.

Mention should now be made of the salient contribution by Francis W. Aston, who had worked on isotopes in the Cavendish Laboratory under J. J. Thomson before World War I. These researches were interrupted by war work, when he studied the chemistry of dopes for aeroplanes. Then in 1919 he invented what is known as the *mass spectrograph*, for which he became famous and later went on to develop a more improved version. He used the mass spectrograph to determine, very precisely, the atomic masses of a number of gaseous elements as well as their isotopes. Aston showed that the masses of individual atoms were invariably a little smaller than the sum of the masses of their constituent particles. It now had to be assumed that some mass must be lost in the process of atom building. So using Albert Einstein's famous equation $E=mc^2$ relating mass to energy, it was possible to calculate the equivalent so-called *binding energy* that holds together the particles which make up the very closely bound entity of the nucleus. The binding energy E, equals m, the mass loss, multiplied by a constant, $c^2$, which is equivalent to the speed of light multiplied by itself. Hence a very large number is obtained for E, the binding energy. This is a simple though very important example of how the work of the theoretical and experimental physicists was coming together in the years before World War II.

## The discovery of fission

In 1934, Irene Curie and Frederic Joliot-Curie jointly discovered that alpha particle bombardment of stable, light elements like beryllium, boron and aluminium could artificially turn them into radioactive elements. Enrico

Fermi and his co-workers then used neutrons to bombard as many as 60 different elements, including some of the heavier ones such as uranium, and also determined their half-lives. Fermi's results attracted the interest of the Austrian physicist Lise Meitner and the German chemist Otto Hahn who confirmed Fermi's findings. Then Hahn and Fritz Strassman, working together in Berlin, did a considerable amount of work in this field which seemed to indicate that the effect of bombarding uranium with neutrons was to produce radioactive isotopes of barium and other elements. The mass of these elements was approximately half that of the original uranium so it began to look as if they were virtually splitting some of the uranium nuclei. Hahn then wrote about this to Meitner, now in England since being driven out of Germany in July 1938 by the Nazi persecution of Jews. Meitner told her nephew Otto Frisch, a physicist, and they managed to explain the phenomena using Bohr's model of the nucleus. They concluded that the colliding neutron set up violent internal motion in the uranium nucleus causing it to split into two more or less equal fragments, as suspected, each having half the mass and half the nuclear electrical charge of the uranium atom. For this process, Frisch suggested the name 'fission' from its similarity to the division of a biological cell. The amount of energy released, the so-called binding energy, was calculated to be very considerable, which meant the fission fragments would fly apart at great speed carrying most of the released energy with them.

The discovery of fission was soon known throughout the world of physics for it was published in two letters which appeared in the scientific journal *Nature* in February 1939. Reports which confirmed the discovery soon came from various countries — Joliot-Curie in France, P. H. Abelson in the USA, Egon Bretscher and his co-workers in England, and by Fermi's team. An early development of the theory of fission came from Bohr, who showed that fission was much more likely to occur in the light isotope of uranium, namely U-235 rather than U-238. Also, the probability of fission was much greater for slow-moving than for fast-moving neutrons. Two days before World War II broke out in September 1939, Bohr and Y. A. Wheeler, his co-worker, jointly published what became known as the classic analysis of the fission phenomenon. This appeared in *Physical Reviews*, Volume 56, 1939.

## The chain reaction and critical mass

When a uranium nucleus fissions as a result of neutron bombardment, some free neutrons are thrown out in addition to the two massive fragments. These are called 'secondary neutrons'. Though their possible ejection was first noted by Hahn and Strassman, the first to establish this experimentally were

Joliot-Curie and his collaborators, Hans Halban and Lew Kowarski. The latter team published this result in *Nature* in 1939. Shortly afterwards, they stated that on average, more than one secondary neutron was emitted per fission. This indicated that in principle, a self-sustaining chain reaction was possible because the secondary neutrons could cause fissions in other uranium nuclei. Provided that more than one secondary neutron was produced per fission, the reaction could spread from atom to atom through a large mass of fissile material at a very fast rate. Each fission of a uranium nucleus results in the release of over one million times as much energy as in the combustion of an atom of carbon, so the energy output rate could be very great indeed.

In practice some neutrons would escape altogether without hitting any uranium nucleus. The greater the mass of material, the less chance there is to escape without collision and we reach what is called the 'critical mass' required in practice to produce a self-sustaining chain reaction. The purity and density of the uranium and the proportion in it of the highly fissile uranium-235 isotope all affect the size of the critical mass, as does the velocity of the neutrons, because escape is more likely for fast than slow neutrons. The notion of critical mass, or size, was first introduced by Francis Perrin in France and followed up by Rudolf Peierls in England. They showed that conditions for a chain reaction could not be established in a lump of refined natural uranium, however large, with fast neutrons. However it was just possible if the neutrons were slowed down. Fermi's earlier discovery now proved useful, for he had shown that when neutrons collided with atoms having about the same mass, they were slowed down in consequence. Thus fast neutrons could be slowed down to the desired speed by allowing them to collide with atoms of a very light element such as hydrogen. This is the scientific basis of the use of a 'moderator' in slowing down neutrons in a nuclear reactor. A hydrogen-bearing material convenient for moderating the speed of neutrons in this way is water, or better still, 'heavy water'. The latter is water in which the hydrogen atoms have been replaced by their isotope deuterium, where the nucleus contains one neutron as well as the normal proton. Much work in this field was carried out in France and a small quantity of heavy water was accumulated for experimental work on chain reactions and moderators.

Obviously, in 1939 the nuclear scientists now had in mind the possibility of making a super bomb or a type of atomic boiler to generate heat/energy in a controlled and useful way. Yet there were still many uncertainties to be taken into account. Indeed the German scientists did not appear to appreciate the difference in effectiveness of slow and fast neutrons with regard to the making of a bomb. Perhaps that was one of the reasons why the Germans did not strive to manufacture an atom bomb during the war, although Britain and America were convinced that they would. But now we have entered the age in which man was about to apply his accumulated knowledge of the atom and its nucleus; the age of nuclear politics, the story of which we begin in the next chapter.

*Note*

In compiling the above chapter, some use was made of the so-called technical chapter entitled 'A Glance at Prehistory', originally prepared by Kenneth Jay at AERE Harwell for inclusion in Margaret Gowing's seminal work, *Independence and Deterrence: Britain and Atomic Energy 1945-52*, published in 1974 whilst she was the UKAEA's official historian.

# Chapter 2

# MAUD and the Manhattan Project

B ritish nuclear politics during World War II mainly centred on relation-ships with the United States. Over the six years a great change took place in the balance between the two countries in terms of economic power and world prestige. By the end of the war, America was to emerge from her pre-war days of isolationism as the dominant and dominating power on the world stage — and the only nuclear power in the world.

## The MAUD Committee

But at the outbreak of war in 1939, by far the greatest concentration in the world of skills and knowledge in nuclear science and technology lay in Britain. Work in Britain on nuclear fission was mainly centred in the univer-sities of Birmingham, Cambridge, Liverpool and Oxford along with ICI research. A number of distinguished nuclear scientists had become aliens overnight, escaping the Nazi persecution of Jews in Germany and Austria. When Germany advanced into France in 1940, two leading French nuclear scientists, Hans Halban and Lew Kowarski, fled to England with their small stock of heavy water. They had to follow a tortuous route from Paris, going first to Clermont-Ferrand then over to Bordeaux where they managed to board the *Broompark*, a British collier lying in the docks. The ship travelled without mishap to England, not only smuggling out the heavy water but also £2.5 million worth of industrial diamonds and some valuable machine tools. When Germany turned on Russia and invaded, that country's nuclear work would possibly have ceased or been considerably reduced, depending on where it was located. However, contact with scientists in the neutral US who were also studying fission might have continued — within the limits laid down by security — until the US were at war with Germany.

In March 1940, two leading nuclear scientists, Otto Frisch and Rudolf Peierls, wrote a memorandum which discussed the possibility of making a nuclear bomb. This was sent to the British government and it produced an immediate reaction; the setting up in April of a committee within the Ministry of Aircraft Production (MAP) to report on the feasibility of Britain producing such a bomb. It was coded the MAUD Committee and it was chaired by Sir George Thomson. It was directed to send its report to the Directorate of Scientific Research at MAP.

At the outset there does not appear to have been any discussion within MAP or the MAUD Committee on the moral aspect of making and then going on to use a nuclear bomb, which was simply regarded as a super-efficient high explosive bomb. It is perhaps surprising that so many alien scientists continued working on the bomb project when, if successful, it could result in a devastating destruction of their homelands and slaughter of many of their fellow citizens, indeed their own friends and relatives. Margaret Gowing, official historian of the British atomic project, suggested that perhaps they were too preoccupied in meeting the scientific challenges and were so deeply committed to the war that such matters did not strike any moral note of conscience. However, Gowing said that someone closely concerned with MAUD had once reflected that 'perhaps we should have studied the moral and political implications of the bomb and thought about its use. Perhaps too we should have considered whether radioactivity was a poison outlawed in spirit by the Geneva Convention. But we didn't.'

By March 1941 there remained only one difficult technical problem before an atomic bomb ceased to be a matter of speculation. Although Sir Geoffrey Taylor had calculated that the critical mass to cause a nuclear explosion with U-235 was only likely to be 8 kilos, maybe less, a method for separating out this isotope from natural uranium had yet to be found. Whichever method was chosen, it would probably be costly to put into practice on the required scale. The favoured approach by MAUD was to turn the solid natural uranium into a gas known as uranium hexafluoride, and then pass this through a fine membrane or gauze. The two isotopes passed through at different rates owing to their difference in mass, thus effecting a degree of separation. The starting material had to be carefully purified and ICI were advancing along this road whilst W. N. Haworth at Birmingham was developing means of purifying the gaseous hexafluoride. Rudolf Peierls and Klaus Fuchs carried out theoretical work necessary to make it possible to design a large-scale diffusion plant and indeed the Oxford team soon had built a small-scale model of one stage of a diffusion plant. In practice, many stages would be required to effect sufficient separation of the two uranium isotopes.

Egon Bretscher and Norman Feather working at the Cavendish Laboratory in Cambridge were following up a different approach to a nuclear bomb. A new element called plutonium had been recently discovered during studies on fission and these researchers were studying its fissile potentialities. Plutonium is not to be found in nature, like uranium, but just conceivably it might be produced in a machine in which a controlled chain reaction with slowed-down, thermal neutrons had been induced. This was to turn out to be a practical possibility and ultimately was the route adopted for obtaining material for the Nagasaki bomb (see Chapter 3). But in 1941 the plutonium work was pushed into the background for it would have involved very heavy capital costs. On the other hand, although the bomb project costs were still quite small up to mid-1941, they were soon likely to rise very considerably in consequence of adopting the diffusion route via uranium hexafluoride.

In July 1941, the MAUD Committee issued a draft report prepared by Thomson. It showed that a plant just to separate one kilo a day of U-235 would cost £5 million (say £100 million in present-day money values). However the committee felt that 'every effort should be made to produce bombs based on this route because of its anticipated destructive effect, both material and moral'. Also it was believed that the Germans were probably working on a bomb of this nature. Continued collaboration with the Americans was thought to be desirable. Dr P. M. S. Blackett, a leading physicist, was a dissenting voice, for he thought a bomb would hardly be ready within two years, by which time the war might well be over. Also, he considered that the full-scale plant ought to be built in America. MAUD also referred to the possible exploitation of fission to provide a machine, or boiler, as the two senior French scientists on the project called it, to release energy in controlled fashion as a substitute for burning coal or oil. These men were particularly enthusiastic about this possibility. Pure graphite blocks or heavy water would moderate the speed of the fast neutrons produced by fission of uranium and make the machine controllable. It was even suggested that ordinary water could act as a moderator provided a degree of enrichment of U-235 in natural uranium had been achieved; the latter plant was thought to be a good deal cheaper than the diffusion plant required to produce bomb material. These possibilities for the boiler route were assessed as an 'outside chance', sufficient to justify keeping Bretscher going at Cambridge.

The report was submitted to the MAP on 29 July 1941. MAUD had by then functioned for 15 months but it never met again. Gowing stated that British work was 'much more effective than American work' at this stage. But of course the Americans were not at war and would hardly be imbued with a sense of dramatic urgency over bomb production, indeed they were probably more concerned with power generation possibilities. Gowing also concluded that without the efforts of MAUD, World War II might well have ended before an atomic bomb had been dropped and 'in this case, the world would have been a different place'. Well, maybe.

After many delays which frustrated Thomson, the report went on in September to Sir John Anderson, Lord President of the Council. He and his advisory council thought strongly that development of the U-bomb should be pressed ahead with all speed. A 20-stage pilot plant for uranium diffusion should be built in Britain but the full-scale plant should be built across the Atlantic for it would occupy several acres of land and there would be a serious risk of air attack over here. The Prime Minister, Winston Churchill, readily approved the proposal to go ahead, being urged on by his scientific adviser, Frederick Lindemann (later to become Lord Cherwell). Anderson was to be the cabinet minister responsible for the project. Churchill insisted that the atomic bomb project be kept very secret and indeed it does not even appear to have been discussed within his highly selective War Cabinet. Clement Attlee, the deputy PM, only heard of it as a 'rather bigger and better bomb'.

So far as power generation went, this section of the report was briefly dealt with. Sir Henry Tizard, an eminent scientist of the day, doubted that uranium power would ever be cheaper than the cost of power from coal, or water-power, although it would take half a century before this came to be widely accepted. Nevertheless it was decided that some development should proceed on these lines and that it should be entrusted to the Department of Scientific and Industrial Research (DSIR). Anderson's advisory panel did not want any chance of business falling into the hands of a 'private commercial interest'. The cover name was Tube Alloys, which was to be run by a Consultative Council that would deal with policy. At the technical level, the atomic power development project would be headed by Sir Wallace Akers, then Head of Research at ICI. The Technical Committee consisted of Professors Chadwick, Simon and Peierls with Drs Halban and Slack, the latter from ICI.

There now followed consultations with the Americans after they had been shown the MAUD Report, which impressed them a great deal. A scientific and technical committee was appointed by the National Academy of Sciences in the US in order to advance work in the nuclear field. It eventually produced a report which was in many respects like MAUD. But there was no real urgency in the US about atomic developments at this time although President Roosevelt endorsed 'complete interchange of information with the British'. But very soon all that would change and the pace of atomic development was to quicken enormously in the US after 7 December 1941 when the Japanese attacked Pearl Harbour and America was at war. Within six months its nuclear weapons effort far outstripped in size and expenditure that of the British and a great sea change began to take place in Anglo-American nuclear relations.

## The Manhattan Project

After Maud was approved Britain had dragged its heels over the question of whether or not to go for a joint bomb project with the Americans, but by the middle of 1942 it was too late. In July, Churchill approved recommendations for collaboration with the US, including the building of pilot plant and full-scale diffusion plants over there, but it never came off. The Americans were already busily building their own and Anglo-American collaboration was breaking down.

Major-General Groves was appointed overall head of US atomic bomb activities which came to be called The Manhattan Project. Security became very tight, a process of 'compartmentalisation' was introduced and only Groves and his top aides had overall knowledge of what was going on. The Americans felt that as they were now responsible for 90 per cent of the Project they should press ahead virtually on their own. They also feared that collaboration might slow things down and both President Roosevelt and Henry Stimson, his Secretary for War, now supported playing down collaboration with Britain. But they failed to spell this out and hence much time and effort was spent by British politicians and scientists in chasing after the Americans. Churchill raised the matter with the President's personal aide at the Casablanca conference and followed this up with urgent telegrams, all of no avail. Privately, the Americans were worried about giving the British a commercial advantage after the war was over and they also had serious doubts about whether Britain should then be allowed to share in the possession of atomic weapons.

Churchill called for the costs of Britain going alone but these proved to be prohibitive, trivial though they would have been compared with the American figures. After three or four years of war Britain was fully stretched with regard to materials, engineering equipment of all kinds and technical manpower. Fuel and power supplies especially had become an acute problem. Britain also lacked engineers of every description. Certainly it would have been impossible to contemplate having an atomic bomb ready before the war had ended. By now, the US government had voted well over £250 million for the atomic project, several billions at present money values, but all explanations as to what was going on had been denied to Congressmen on security grounds. It was no wonder the President's office was cool about having Britain involved in the Manhattan Project, even though several nuclear scientists from this side of the Atlantic were now working over there.

Nevertheless the President and the Prime Minister did eventually come to a form of understanding and agreed on some degree of collaboration. Hence they signed the Quebec Agreement (QA) on 19 August 1943. This acknowledged that 'it would be improvident use of war resources to duplicate plants on a large scale on both sides of the Atlantic and therefore a far greater expense has fallen upon the United States.' The agreement referred again to the heavy burden of production falling upon the US and went on to say that Britain accepted that 'any post-war advantages of an industrial or commercial character shall be dealt with as between US and Great Britain on terms specified by the President of the United States to the Prime Minister of Great Britain.' And so the Americans intended to call the shots after the war regarding the power applications of nuclear energy but there was nothing to imply they would prevent Britain starting to make its own nuclear weapons.

There were other matters specified in the QA, mainly concerning sharing of information and decision-making. The two countries agreed never to use the agency against each other, nor to use it against a third party without each other's consent, nor give away any information to a third party without mutual consent.

Margaret Gowing, whilst official historian for the United Kingdom Atomic Energy Authority during the 1960s, stated that the QA brought to an end a long and unhappy episode in Anglo-American relations. She said it was 'the fruit of a basic misunderstanding and lack of generosity in both British and Americans.' This is too balanced a view. More important is the fact that the QA was a clear pointer to the weakening of British influence in its relations with the United States, a result of Britain's decline in economic power. The US had now left behind its old policy of isolationism to become the dominant power of the so-called Western Alliance.

By the time of the signing of the Quebec Agreement in August 1943, the Manhattan Project was in full swing. It was by now evident that the entire US programme was to cost over $1000 million, all on the bomb and nothing on long-range possibilities of nuclear energy for power production. Of course much of the developing technology would have applications in nuclear power generation, as indeed would much of the more basic nuclear scientific research being carried out.

In addition to large-scale diffusion and other types of plant for separating the U-235 isotope from natural uranium, aimed at producing a uranium bomb, a full-scale graphite-moderated pile to produce material for a pluto-nium bomb was well under way. It was anticipated that materials would be ready for a first weapon test early in 1945.

James Chadwick, later Sir James, head of the British mission, acknowl-edged the logic of Britain's part in the Manhattan Project. It was to assist the Americans 'to the utmost of our ability' although he fully accepted that the success of the Project would not depend on British involvement. He actually thought that Britain was fortunate to be allowed to participate and it was to our advantage to have as many scientists in America as possible. According to Gowing, the latter policy virtually resulted in the closing down of the UK atomic project.

But this is to overlook the Anglo-Canadian developments in a nuclear laboratory at Montreal in Canada. Chemists and physicists were at work there in 1943. They were hoping to have a nuclear pile in which a self-sustaining chain reaction in natural uranium would result from the moderation of neutron energy by using heavy water. At the end of the year, Chadwick managed to persuade Groves to agree to the building of a large-scale heavy water pile at Montreal for the purpose of producing plutonium as bomb material. The pile would have a power of 50,000 kilowatts and was coded 'NRX'. The site chosen was Chalk River on a bank of the Ottawa river. The Canadians agreed to bear the full capital costs, after all it would be on their territory, but they would not be paying the salaries of the British staff. By September 1944, over 40 Canadian professional staff were attached to the Montreal laboratory along with an equal number of British experts. In addition there were 12 scientists of other nationalities including Halban and Kowarski who had fled to Britain from France in 1940, plus a small group of young New Zealanders.

Although it was hoped to commission NRX in a year or so, this proved impossible and it was not accomplished until July 1947, two years after the war had ended. It had been designated by Major General Groves as 'a pile to produce plutonium for nuclear bombs for use in World War Two', so why did not Britain and Canada move to negotiate abandonment of the NRX project when the war ended and the Hiroshima and Nagasaki bombs had been exploded? Britain had voted for a post-war Labour government without any reference to a nuclear weapons project. There was a desperate shortage of scientific and technical people at that time and the Montreal project could no longer claim priority over home needs. There was a good case for bringing home British members of the Canadian project to their families in 1945 rather than keeping them out on a redundant war project.

In 1944 the Montreal Project had become headed by John D. Cockcroft, later to become Sir John Cockcroft, head of the Harwell nuclear research station back in England. Indeed some of the studies at Montreal were of a forward-looking nature and this work was concentrated in what was called The Future Systems Group. Many possible types of pile, or nuclear reactor, were considered. Chadwick thought that an 'adequate' stock of pure fissile material suitable for military purposes should be accumulated. The people out in Canada were in fact implementing a nuclear policy for post-war Britain and it seemed to be starting from an assumption that we were going to make our own nuclear weapons after World War II (see Chapter 4).

# Chapter 3

# Hiroshima

---

After going to America in 1943, Niels Bohr, the Danish world-renowned theoretical physicist, became increasingly concerned about the political implications of the atomic bomb for post-war international relations. He was alarmed to hear scientists at the Los Alamos laboratory discussing the ultimate possibility of a hydrogen bomb being developed. This would be based on the fusion of atomic nuclei and its potential explosive power would be enormously greater than that of the atomic bomb currently under development.

## Scientists try to avert an arms race

Bohr voiced his concerns to senior politicians including Sir John Anderson who, despite earlier enthusiasm for the bomb, had forebodings about a post-war nuclear arms race between two competing power blocs and who feared the wider effects of the bomb on international relations. Lord Halifax, the British Ambassador in Washington, was similarly concerned about the future.

By 1944, Bohr had come to feel that it would be disastrous if Russia were to find out on her own about the USA/UK bomb project. Lord Halifax felt the best approach would be to President Roosevelt in the first instance and Bohr therefore did so, via a Mr Justice Frankfurter who was a supreme court judge and an old acquaintance of Bohr. The message which came back from the President was encouraging; he said the whole business 'worried him to death' and he was 'most eager to explore proper safeguards with Mr Winston Churchill.'

In the meantime, back in England, Anderson consulted Lord Cherwell (formerly Professor Frederick Lindemann), on whom Churchill normally leaned heavily for advice on technical and scientific matters. Anderson then sent a long minute to the Prime Minister, suggesting it was high time to inform the War Cabinet and Chiefs of Staff about the present position on the atomic project and its implications. Anderson said there were only two alternatives facing Britain;

(a) a nuclear arms race, with the US and UK having a precarious advantage at the outset; or

(b) some sort of international control.

Anderson favoured (b) but realised that the PM could not make a move without the Americans. However, he believed a few important people around the White House were becoming increasingly concerned about the future. Finally, if the PM were to adopt option (b), then there was much to be said for communicating to Russia that an atomic bomb was well on its way to being prepared. He thought the Russians should be invited to collaborate in preparing a scheme for international control.

Churchill tersely wrote on the end of Anderson's minute; 'I do not agree.' He was adamantly opposed to any breach in the wall of secrecy surrounding the bomb project. He would not even consult Clement Attlee, the deputy PM. Anderson then had another go; he sent Churchill the draft of a possible telegram to go to President Roosevelt. This also received a curt reply: 'I do not think any such telegram is necessary.'

Then Lord Cherwell tackled Churchill direct on the subject of international control, following a talk he had had with Mackenzie King, the Canadian Prime Minister, who believed the situation now required discussion. But once again, Churchill refused to move his position.

It was now the turn of Sir Henry Dale, the President of the Royal Society, to approach Churchill. In a letter he said; 'I cannot avoid the conviction that science is approaching the realisation of a project which may bring either disaster or benefit on a scale unimaginable, to the future of mankind.' But it was of no avail.

Bohr now became keen to see Churchill himself and this he did in company with Cherwell, in the May. As might have been expected the meeting was a failure. Bohr came away quite depressed at the way the world was apparently governed with 'small points exercising a quite irrational influence' as he saw it. 'We did not speak the same language,' he concluded. Then finally, Field Marshall Smuts of South Africa wrote to Churchill in June on the subject of international control but of course nothing came of this contact either.

Whilst Churchill remained stubbornly opposed to any action being taken, across the Atlantic there was a promising meeting between Roosevelt and Bohr in complete privacy on 26 August. Roosevelt said on this occasion that an approach to Russia was vital and it could open a new era in human history. He thought Stalin was a sufficient realist to understand the implications of this scientific revolution. But Roosevelt must have realised the difficulty of getting the British PM to act. He even told Bohr how, at the Teheran Conference, he had made jokes to prevent clashes between Churchill and Stalin! Bohr was so encouraged by his meeting with Roosevelt that he followed this up with a memorandum on how the contact could be made with Russia via the scientists and he even had a shot at writing a draft of a letter to Peter Kapitza, a leading Russian scientist well known to Bohr.

But Bohr's hopes were dashed when Churchill met Roosevelt in September at the second Quebec Conference. Here, the President did a complete volte face and along with Churchill, signed the aide memoire which specified that no information on the bomb was to be communicated to any other country. The bomb 'might perhaps, after mature consideration, be used against the Japanese, who should be warned that bombardment would be repeated until they surrender.' As regards 'mature consideration', in the event, largely owing to Churchill's stubborn behaviour, the British did not get involved. Gowing stated that in studying all the relevant documents she could find no reason for Roosevelt's complete change of attitude on meeting Churchill.

A dejected Bohr decided to abandon his attempts at persuasion and went to spend the rest of 1944 at the project's Los Alamos laboratory. In the December, Einstein, being disturbed at the prospect of an atomic arms race after the war, contacted Bohr to propose that scientists got together internationally to discuss the situation. He named Bohr (Denmark), Einstein and Compton (US), Lindemann (Britain), Kapitza and Joffe (Russia). Bohr advised Einstein sadly that the action he had proposed was 'illegitimate' and eventually Einstein agreed to abandon the idea.

Bohr visited England again in March 1945, where he found Churchill was still vehemently opposed to any approach to Russia, or even to the French; it will be recalled that the Germans had been driven out of France by the previous summer. Bohr was now growing desperate because he knew the bomb would very soon be ready and on returning to the US in April, he got together with Halifax and Frankfurter once more with a view to making a further approach to Roosevelt. As they were ending their discussion they heard the bells ringing out in Washington — the President had died. However Bohr did not give up but sent a memorandum to Henry Stimson, still Secretary for War. This was followed up by Halifax and Frankfurter who pressed Stimson to take action and this he did. He set up what was called the Interim Committee and took the chair in person.

The committee was divided in its views, with the scientists pressing for an immediate approach to Russia, China and France before the bomb was used and the others opposing this. At the end of June they came to a compromise;

to urge Truman, the new president, to inform the Russians at the forthcoming Potsdam Conference about the existence of the atomic bomb and also say that the US anticipated using it on the Japanese, but that it was hoped that future discussions would take place to ensure the new weapon became 'an aid for peace'.

The British took no part in the Interim Committee deliberations but at the Manhattan Project's Combined Policy Committee (CPC) meeting in early July, they did hear of the final decision to disclose to Russia. When subsequently Stalin heard the news at Potsdam he seemed strangely casual about it.

## Hiroshima

The decision to use the bomb against the Japanese without warning had been taken at the June committee meeting (ironically, it was the same month in which the United Nations was founded) and when Churchill saw the minute formally asking for his agreement he merely initialled it. So 4 July saw the CPC in Washington duly noting British agreement. Obviously Churchill never expected to be involved in making the decision, he anticipated correctly what it would be and it was presumably what he wanted.

The first bomb was now ready and was exploded on 16 July at Alamogordo in the Mexican desert. The force of the blast exceeded all expectations.

The Interim Committee continued its deliberations through the summer of 1945. Apart from discussing the wider implications of the bomb they also seem to have agonised over how to use it against Japan. Then a powerful, well thought out memorandum was received from a group of senior scientists working in the project at Chicago. Their spokesman was one James Franck and the memorandum was published in full after the war under the title of the Franck Report and it was dated 11 June 1945.

The various considerations set out in the Franck Report led the authors to conclude that the use of nuclear weapons against Japan, unannounced, was not advisable:

'It would precipitate an arms race and prejudice against reaching an international agreement on the future control of such weapons. More favourable conditions for eventual agreement would be possible if the bomb was first revealed to the world by demonstration in an appropriately selected uninhabited area.'

The report urged that consideration of the use of nuclear bombs be weighted in favour of long-range national policy rather than military expediency. Finally it said international control was vital for the USA because the only effective method of protecting the country appeared to be the dispersal of its major cities and essential industries.

Needless to say, the advice of the Franck Report was ignored. There was a strong feeling at the Chicago site about the use of the bomb and a poll held among 150 scientists gave the following result:

46 per cent were for a demonstration in Japan and then a chance to surrender
26 per cent for an experimental demonstration in USA in front of Japanese representatives
15 per cent for using atom bombs in the most effective way militarily
11 per cent for a public demonstration and no further action
2 per cent for keeping the bomb secret and foregoing combat use

Anyway, the die was cast. The first atomic bomb used against Japan was exploded over Hiroshima at 8 o'clock on 6 August. Early assessment of the resulting deaths had put the figure at about 64,000 within 4 months but later investigation, taking account of cancer and related deaths resulting from the bomb, puts it nearer to 200,000. There were over 72,000 injured and about 4 square miles of the city utterly destroyed with a further 9 badly damaged. On 8 August, Russia declared war on Japan and invaded Manchuria. The second bomb was used on Nagasaki on 9 August, this causing at least 39,000 immediate deaths and 25,000 injuries. The first bomb was based on U-235 and the second one was a plutonium bomb. It was originally intended to drop the bomb on Nagasaki on the eleventh but this was brought forward by two days.

Many books have been written about all aspects of these two bombs, still the only nuclear weapons to have been exploded in anger. No further space is therefore taken up here except to say that no justification has ever been given for dropping the second bomb. Was it to prevent Russia gaining influence in the Far East, having just entered the war against Japan? But then the decision to drop both bombs must have been taken before it was known that Russia was to enter the war.

As for Bohr, with the establishment of Stimson's Interim Committee he felt he could do no more. In June his wife had come over from Denmark to England and he flew to join her and return with the united family to Copenhagen. His self-appointed wartime mission had not succeeded inasmuch as there was no consultation outside the US/UK as regards possible plans for future international control of the bomb. But until his death, anxieties over atomic weapons and international control over them remained his prime concern.

It is possible in retrospect to conjecture that the talks which Bohr was working for could possibly have begun as far back as 1944 when Roosevelt was alive and might even have made headway had it not been for one stubborn man — Prime Minister Churchill.

# Chapter 4

# Attlee's secret nuclear bomb project

In April 1945, the month in which Adolf Hitler died in Berlin, the coalition government under Winston Churchill decided to undertake a broad programme of research and development in atomic energy, this to include the building of what was to be called the Atomic Energy Research Establishment (AERE). In July, John D. Cockcroft was designated director of this establishment although as we saw in Chapter 2, he was in Canada at the time, heading up the project centred on the NRX reactor. On 26 July it was announced that Labour had won a general election with an overall majority of 146 seats and thus Clement Attlee became Prime Minister. As described in the previous chapter, the atomic bombs fell on Japanese cities on the 6th and 9th of August, with the approval of Churchill on behalf of Britain whilst he was still Prime Minister. This raises the question as to how much Attlee knew about the plan to use nuclear weapons on densely populated cities. Theoretically he would just have had time to send a protest to the Americans on his return from the Potsdam Conference, thus negating Churchill's approval, had he so wished. But the probability is that he had been kept unaware of the decision; it is a fact that Churchill communicated very few decisions concerned with the war to his deputy PM.

After Labour took over the reins of government in the summer of 1945 it would have been open and democratic to have made information available to enable an informed general public discussion to take place over what to do about nuclear energy. Perhaps this would have resulted in the options being put to the electorate — a referendum, in fact. But nothing of the kind took place. There was no stimulation of interest, and nuclear matters were kept under wraps. Certain eminent scientists knew about the hidden agenda and some were opposed to what they feared the government was about to do. The leading wartime physicist P. M. S. Blackett (later Lord Blackett), instigator of operational research in aid of the Royal Navy during the war, wrote in a confidential memorandum sent to the cabinet in November 1945:

'To make [nuclear] bombs or acquire them now would tend to decrease rather than increase our long-term security. If we needed them, immediately, for our security, then we should ask the US for a supply.'

He went on to make clear that our policy should be not to make atomic bombs. Blackett believed there were grave disadvantages for any small country like Britain participating in a nuclear arms race in the post-war world. The new PM and the Chiefs of Staff disagreed with him. But then, Blackett's war-time ideas for reducing our loss of merchant vessels at sea had initially been rejected by the Admiralty as being crazy. However the admirals did eventually put his ideas into operational use, resulting in a significant reduction in loss of shipping.

Two other top scientists, Sir George Thomson and Sir Henry Tizard, thought the development of atomic energy should be put off for some time, this in the interests of much greater security. Interestingly, they also believed that the possibilities of peaceful atomic energy exploitation were being much exaggerated. It was unlikely that electricity could be produced any cheaper from a steam powered station based on fission than from one based on coal and Britain had coal reserves which would last for 200 years. But several decades were to pass before this would become generally accepted.

A week before Christmas 1945, an event took place in 10 Downing Street of great significance for the future of Britain. A top-secret cabinet committee (coded GEN 75) met under Attlee and took a momentous decision. Only two members of the cabinet were present; Hugh Dalton, the Chancellor, and Stafford Cripps, the President of the Board of Trade. The decision was to build an atomic pile 'as a matter of urgency', this to be large enough to produce material equivalent to 15 nuclear bombs a year. Although GEN 75 does not appear to have specified the aim of the programme, the mere fact of quoting the pile capacity in terms of the number of bombs and the reference to 'a matter of urgency', implied the objective was to manufacture atomic bombs. The close secrecy surrounding the project pointed to this. In the following month, January 1946, the Chiefs of Staff sent a memorandum to Attlee on Britain's atomic bomb requirements. In the August, the Air Ministry actually sent a formal requisition to the Ministry of Supply for an atomic bomb. The die was cast.

## The programme gets under way

A decision had already been taken by the coalition government to go for a nuclear research and development establishment after the war. This was casually referred to at the end of the Churchill administration as 'a piece of unfinished business left for the incoming Labour government to complete'. It was in fact completed in October 1945 when Sir John Anderson, chairman of the Advisory Committee on Atomic Energy, put it to Attlee to formally endorse. This he duly did without hesitation. The site chosen was RAF Harwell in Berkshire. It was a relatively new airfield and was only given up by the Air Ministry on the insistence of the PM. One wonders what arguments were used to convince the Air Marshalls of the vital necessity for them giving up their airfield. Harwell's terms of reference were beautifully vague — 'to pursue all aspects of the use of atomic energy'. It was to be a large establishment and would cost a great deal of money.

But when Attlee's GEN 75 committee decided at the end of the year to go in for building a large plutonium-producing pile and associated complex chemical plants, it brought the government into the realms of really big spending. The pile, shortly to become two, was to be of the graphite-moderated type and built on the site of a Royal Ordnance factory known as Windscale Works, on the west coast of Cumberland (now known as 'Sellafield', Cumbria). For several years the purpose of these piles was kept deliberately vague, indeed they could have been regarded as development piles, built for experimental purposes. In February 1946, Christopher Hinton (later Lord Hinton) was appointed head of design, construction and operation for the plants at Windscale. He was a very experienced manager of large engineering construction projects having been Deputy Director of Filling Factories for the Ministry of Supply (MOS) during the war. But his empire was destined to grow beyond the bounds of Windscale, to include a site near Preston where the uranium for the Windscale pile fuel elements was to be processed.

The Windscale plants would only produce material for plutonium bombs, but soon the government would decide to make U-235 bombs also, although it is not clear why they should have taken this on. A large-scale diffusion plant to produce this material from natural uranium was to be built at great cost and the consumption of electric power would be considerable. Hence a suitable location on the electricity grid was selected and this turned out to be at the village of Capenhurst in the Wirral peninsula. This plant also became Hinton's responsibility.

Since these various production sites would only produce the material requirements, William Penney, a senior physicist, was made responsible for the actual development, fabrication and assembly of the nuclear weapons. He was already Chief Superintendent of Armament Research and highly experienced in that area of work. His headquarters were to be in the Berkshire countryside near the village of Aldermaston, in which area the atomic weapon fabrication plants were to be built.

Attlee eventually realised that a formal decision to manufacture nuclear weapons would have to be recorded and a new ad hoc secret cabinet committee, GEN 163, met in January 1947 to do just that. Present were well-known Labour leaders such as Herbert Morrison, leader of the House of Commons; Ernest Bevin, foreign secretary; and A. V. Alexander, defence minister. Attlee would have informed them that the meeting was in the nature of a rubber stamping exercise since the project was already under way. He was strongly supported by Bevin. No one proposed a referral back for full and detailed costings; they all seem to have endorsed the project without question. And even though the manifesto on which the government had fought the recent election had not included an atomic weapons programme the matter was not referred back to the parliamentary Labour Party for discussion, let alone put before the Commons.

## Parliament is informed

By the middle of 1948, this massive programme was getting well under way yet still nothing had been said in parliament or to the media. But they had to be told and it was slipped in by the Minister of Defence as unostentatiously as possible in answer to a parliamentary question on 12 May 1948. The record reads thus:

> Mr Geo. Jeger: Is the Minister of Defence satisfied that adequate progress is being made in developing the most modern types of weapon?

> Mr A. V. Alexander, Minister of Defence: Yes, Sir. As was made clear in the Statement relating to Defence 1948 (Command 7327), Research and Development continue to receive the highest priority in the defence field and all types, including atomic weapons, are being developed.

> Mr Jeger: Can the minister give further information on the development of atomic weapons?

> Mr Alexander: No, I do not think it would be in the public interest to do that.

> HOC Deb., Vol 450, col. 2117.

These few lines of print were the only information given out to the public after two-and-a- half years in which Britain had been secretly engaged in becoming a nuclear military power. And it went almost unnoticed because there had been a communist coup in Czechoslovakia earlier and then in June, shortly after Mr Alexander's statement, Russia began its blockade of West Berlin. Defence was getting top priority in the Labour administration.

## The nuclear bureaucracy

Despite intense security, or perhaps because of it, the number of committees and subcommittees concerned with Britain's atomic energy project steadily grew.

The Ministry of Supply (MOS) had overall responsibility for the atomic programme but Lord Portal, former Chief of the Air Staff, was in charge of the Department of Atomic Energy (D At En). He had been appointed by Attlee, to whom he was directly responsible rather than to the Minister of Supply. This was presumably as a result of Attlee's fanatical preoccupation with atomic secrecy. Thus Portal did not report to his own Minister and instead had direct access to the PM. According to Margaret Gowing, he was considered to be 'of elevated status and above debate'. This was crazy.

The Ministry of Defence (MOD) had been set up towards the end of 1946 and Tizard was appointed Chief Scientific Adviser. He was also made chairman of the new Defence Research Policy Committee (DRPC). But although Tizard was to advise on 'matters connected with the formulation of scientific policy', atomic energy seemed to have been excluded. This was of course quite illogical. The DRPC set up its own sub-committee to explore the strategic aspects of atomic energy. Attlee seemed to think that Tizard received information on atomic matters, but this was just not so, for in June 1947 Tizard stated at a meeting of defence Chiefs of Staff that no decision had been taken to produce nuclear bombs whereas they knew full well that he was wrong. The Chiefs of Staff had indeed followed up GEN 75's decision of December 1945 with a report, in the following month, on just what atomic bombs were required. In August 1946, the Air Ministry had gone so far as to send an official requisition for atomic bombs direct to the Ministry of Supply.

By the end of 1947, Tizard was becoming increasingly irritated by his own inability to enter effectively into the atomic arena and this eventually led to a clash with Portal. It was becoming apparent, though, that the latter's own position was weakening within the MOS. He soon became part-time and eventually left. Attlee appointed General Sir Frederick Morgan to succeed Portal. It was generally believed that the PM had wanted Sir William Morgan but had got the two names confused. How could the civil service make such a mess of it? Perhaps it was because the PM's tightness over security had left staff improperly briefed. In the event, Morgan behaved like a mere figurehead, his coordination and control was weak and when the Labour government departed from office in the autumn of 1951, Lord Cherwell was brought in by Churchill to fill the post.

Expenditure on the atomic programme was concealed from parliament by burying it in general sub-heads of the MOS vote. The Select Committee on Estimates was not too happy about this undemocratic procedure but when Attlee handed over the premiership, he was congratulated by Churchill on how well he had done in hiding matters of such scale and import from parliament and thus from the whole electorate. During the post-war Labour government there was never a debate in Parliament concerned with Britain's atomic energy programme.

The price of this special status of the atomic energy programme, of its labyrinthine committee system and its excessive secrecy, was confusion in some quarters and ignorance in others, all of which reduced efficiency. According to Gowing, it was the competence and leadership of the three key men at the working level — Cockcroft, Hinton and Penney — that prevented the British nuclear project from being 'an expensive fiasco'. But the nuclear project to make a bomb was a fiasco from the outset, despite the efforts of these able men. The blame lay at the top, with Attlee and his inner ring of Ministers and Chiefs of Staff, for starting off an unnecessary project with an in-built capacity for consuming very large sums of money and quantities of material resources when the country so desperately needed them for post-war construction, re-housing, consumer goods and hardware of every description. Furthermore, the imposed blanket of secrecy was going to prevent any public audit or control of the project's finances.

All governments make decisions at times which, with hindsight, do not turn out to be in the best interests of their country. But except in wartime when Emergency Powers Acts are in force, all projects involving huge, ongoing expenditure out of the public purse require a wide consultation of experts and civil servants, followed by parliamentary debate. This all takes time. But it is democracy at work. Attlee and his ministers responsible for the nuclear weapon decision-making behaved as if they had never heard of democracy.

It is now over half a century since Britain, on her own and in peacetime, made the decision to become a nuclear power. Subsequent administrations, both Labour and Conservative, were all keen to pour public money into nuclear activities. The very fact of this ongoing cross-bench agreement does not of itself prove that what governments were doing was ethical, wise, made economic sense or was essential for Britain's defence. Many billions of pounds were poured into nuclear weaponry and nuclear electricity over succeeding

decades. Although the layman was given to understand that electricity generated from nuclear energy was going to be very much cheaper than that from
coal or oil-fired power stations, in practice it turned out to be more costly, even
after fifty years of experience. Furthermore, as we shall see later in the book,
the two main routes for exploiting nuclear fission — producing nuclear
weapons and generating electricity for civil use — turned out to be inextri-
cably intertwined.

It was easy enough for a Prime Minister and one or two senior cabinet
ministers to get started in the nuclear business, but years later it would turn
out to be much more difficult to escape from it. Britain would in the end be
left with what may be called the nuclear legacy (see Chapter 25). It would be
impossible to ignore highly radioactive nuclear stations at the end of their
useful life, or laid-up nuclear submarines, to say nothing of dealing with the
radioactive and medium active nuclear wastes, the spent nuclear fuel elements
and obsolete nuclear weaponry. And the longer Britain goes on making
nuclear weapons, processes nuclear materials at Sellafield and continues to
operate nuclear electric power stations, the more will the radioactive residues
continue to pile up. Succeeding generations will be left with the costs and
health and safety risks of dealing with the dismantling and decommissioning
of plants and weaponry along with the storing and disposal of radioactive
residues. Some of the most dangerous radioactive materials will remain so for
hundreds of years and although many billions of pounds have already been
spent on the nuclear business, more billions will have to be committed for the
purposes of clearing up.

The world was at peace in 1945 but suffering exhaustion after six years of
world war. Economically, Britain was almost ruined. There was a housing
crisis and rationing of major commodities, noticeably foodstuffs. There were
serious shortages of many building and engineering materials as well as
durable household goods. In the face of so many problems, heads of departments and cabinet ministers hardly knew where to turn. But the ministers in
the know on the atomic weapons programme should have realised that such
a costly programme would divert already scarce materials away from the
urgent needs of the domestic front.

Why was there such intense secrecy surrounding the nuclear decision by
three top ministers of a Labour government, only six months after winning
the post-war election with a large working majority? No effective pressure was
to be expected in parliament from the opposition. It may have been to prevent
the Soviets learning about it and also the Americans. It could well have been
to maintain British prestige in the world; but it could not achieve this if it was

kept a secret! Some on the left, keen to preserve Attlee's reputation, have said that he must have been under great pressure to come to such a decision, probably from the Chiefs of Staff. But it may simply have been the result of the momentum generated by Britain's involvement in this new technology during World War II — the NRX project was still in progress in Canada — together with a belief that the military purposes could not be separated from the potential to harness the power of the atom for peaceful purposes. The true reason for Attlee making his historic decision has never been disclosed and is likely to remain in top secret archives for a long time to come.

# Chapter 5

# The UN Atomic Energy Commission, the McMahon Act, and the Cold War

In San Francisco on 26 June 1945 the world's statesmen signed a charter to establish an organisation whose prime purpose was 'to save succeeding generations from the scourge of war'. This signing took place only a matter of days after the Americans had made the decision to drop atom bombs on Japan without prior warning. When six weeks later the bombs fell on Hiroshima and Nagasaki, people the world over were horror-stricken.

The first UN General Assembly met in London on 24 January 1946 and its very first resolution, numbered 1(I), was to set up an Atomic Energy Commission (UNAEC) charged with preparing proposals for 'the elimination from national armaments of atomic weapons...' This resolution was unanimously accepted by all 51 member states. What an auspicious beginning it must have seemed, even though all present would have known that the US already possessed atomic weapons and had used them in war. However the General Assembly would have been unaware that the British Prime Minister had one month previously made a secret decision to start manufacturing nuclear weapons. We must also assume that the Soviet Union was by now working secretly all out to produce such weapons, otherwise they could not possibly have been in a position to explode one only three years later.

The US now prepared a plan for the control of nuclear weapons to which the leading American nuclear scientist Robert Oppenheimer made a major contribution. It was initially called the Acheson-Lilienthal Report. It called for the UNAEC to have real power and recommended unconditional American co-operation. It accepted that the US currently had an extremely strong position regarding atomic devices but it would not last — 'we must use this advantage now to promote international security'. The American president chose Bernard Baruch, a financier, to translate the report's proposals into terms more favourable to the US and his revised report — The Baruch Plan — was presented to the first meeting of the UNAEC on 14 June 1946. Its underlying principle of control preceding prohibition of weapons, together with threats of punishment and demands for inspections associated with control, raised Russian suspicions; they refused to accept the Plan and therefore exercised their right of veto. Attlee must have sighed with relief, for otherwise he might well have decided to exercise the veto and would have risked blowing his cover.

The Soviet delegate, Andrei Gromyko, now countered with a Soviet plan of their own, which was to destroy all existing atomic weapons and cease all production of them. But the Americans would not countenance this plan since it was founded on the principle of prohibition preceding control and they somewhat rudely countered it two weeks later by carrying out the world's fourth nuclear explosion, over Bikini Atoll, before the UNAEC Technical Committee had time to begin work.

Britain's Advisory Committee on Atomic Energy, though unhappy about the Baruch Plan, yet failed to prepare a scheme of its own and the British showed no initiative at the UNAEC. Attlee and his close aides were of course apprehensive of any scheme likely to cause Britain to give up its own newly-founded nuclear programme and hence the Soviet plan was rejected. The Americans thought Britain was not giving the UNAEC the attention and importance which it merited. Indeed, they were right; the very idea of international inspection of nuclear sites horrified the Chiefs of Staff, even if other governments would be prepared to accept it.

If the British were uneasy with the Baruch Plan the Soviets regarded it as dangerous. The Americans already had nuclear weapons and would retain them for a long while, whereas the Russians would be stopped from acquiring them, and so they were being asked to accept a basic inequality of power. The British physicist P. M. S. Blackett also saw that the Russians would be in a dangerous situation if they accepted the Plan for it would be interpreted in the US as an admission of weakness and might encourage them to start a preventive war. Indeed, both Blackett and Sir James Chadwick had found this very idea actually being expressed within the US at that time, and by people who had normally held liberal views with regard to international affairs.

There was no significant chance of agreement on nuclear weapons at the United Nations until 1955, as we shall see in Chapter 7, after Nikita Khrushchev had come to power in Russia.

## The McMahon Act

For two or three years after the war, relations between Britain and the US were decidedly cool, to say the least. At the end of 1945, as we saw in Chapter 4,

Attlee was about to begin planning for a nuclear weapons programme while simultaneously trying to cope with a dire economic crisis. The Americans were appealed to for help over the latter situation and they made Britain a loan of $3.75 billion. This action by the US government was unpopular with Congress and so it was granted on very onerous terms. Churchill, along with 70 Conservative and 23 Labour MPs, abstained on the motion to accept the loan. If Parliament and Congress had known that Attlee was about to embark on a costly atomic weapons programme it is unlikely that he would have got his dollars! A year later, Britain was struggling with the economy of her occupied zone of Germany, which was proving too expensive in dollars, and so she suffered the humiliation of having to combine with the US zone.

On nuclear information exchange, the Americans were proving difficult. Although in November 1945, Harry S. Truman (USA), Clement Attlee (UK) and Mackenzie King (Canada) had agreed on 'full and effective collaboration on atomic matters', this was swept aside in 1946 by the passing in America of the McMahon Act which deliberately restricted the flow of nuclear information outside the US. Britain felt she needed all the information she could get on nuclear technology, being in the process of establishing her own programme, and felt the Americans should be generous in this matter. But the McMahon Committee had not accepted that Britain had any 'special relationship' with the US.

Not all the senior British nuclear scientists felt unduly discomfited by this Act. Chadwick asked, 'Are we so helpless that we can do nothing without the Americans?' Hinton took a similar line and thought it would be good for us to 'think for ourselves'. Blackett favoured a neutralist policy, this as part of a neutralist foreign and defence policy. He also thought that we should not be going to make atomic weapons, as noted in Chapter 4. In this he seemed to have support from Tizard who felt that abdication from the military aspects of nuclear energy was the best policy. He stated: 'We are not a Great Power and never will be again. We are a Great Nation but if we try to continue to behave like a Great Power we shall soon cease to be a Great Nation.' And he also said that our task should be not to win the next war but to prevent it from happening.

## Cold War origins

Europe's post-war economic recovery took place against a background of worsening relations between the Western powers and the Soviet Union. Wartime cooperation gave way to conflict, confrontation and what eventually became called the Cold War. Where responsibility for this deterioration actually lies is probably still a very controversial issue in modern history. There had been a legacy of mutual fear and suspicion between the Soviet Union and Western powers ever since 1917, this only broken during the alliance of World War II. After the war, the Russians were determined to guarantee their western borders by means of a buffer zone of friendly states and they were resolved to prevent any resurgence of German militarism. The US was strongly resolved not to accept Soviet hegemony over half Europe or to see its populations forced to adopt Soviet-style political and economic systems. Out of these conflicting basic elements arose a series of events which in effect divided Europe into two hostile camps. The US ceased in 1945 to transmit any more aid to the Soviet Union. The Truman Doctrine in 1947 proclaimed support for 'free peoples resisting attempted subjugation by armed minorities or by outside pressures'. Later in the year, the Russians issued the Cominform Declaration which hardened up the doctrine of two conflicting camps. In February 1948, there was a communist coup in Czechoslovakia which had Soviet backing and the British Foreign Secretary, Bevin, then said he saw no hope of attaining a satisfactory settlement by the Big Four (US, USSR, France and the UK) or through the United Nations.

In the summer of 1948, a crisis arose over Berlin. The city was isolated in the Soviet occupied zone of Germany although all four of the occupying countries shared in its occupation. As a result of a disagreement, Russia attempted to blockade West Berlin, thus cutting off supplies by road and rail to the American, French and British staffs. However, a successful Western airlift kept Berlin alive. But it might be said that this was the point at which the Cold War began. Defence expenditure increased in Britain because of growing fears of war and the length of conscription was renewed at 18 months. In the following April, Britain joined the US, Canada, Iceland and seven countries of mainland Europe in forming the defensive alliance against the Soviet Union known as the North Atlantic Treaty Organisation (NATO). This was immediately followed by the Chancellor's budget speech in which he warned Britain of the growing dollar deficit, pointing out how this was being driven by defence spending. Nevertheless the NATO treaty was approved in the Commons by 333 votes to 6.

As the Berlin airlift began, the US was given permission to station strategic bombers in the UK. Presumably the British Prime Minister felt obliged to accede to the American President's request because of the parlous state of the economy and the ever-present risk of having to be baled out by the Americans. Or perhaps it was a sweetener for getting more nuclear information? Whatever the reason it was a dangerous action to take because in the previous January, Britain and the US had agreed on what was called a *modus vivendi*. Ostensibly, it made for a softening of the McMahon Act so far as Britain went and was a general declaration of intent towards a more formal agreement. American law allowed the President to conclude such an agreement without reference to Congress and it did not come within Article 102 of the UN Charter. But wrapped up in the modus vivendi was a trap — Britain surrendered its veto on American use of the atomic bomb, without even substituting the word 'consultation' for 'consent', a possibility discussed during earlier Anglo-American negotiations on nuclear cooperation. So now we had American bombers on British soil, with the possibility of being armed with nuclear bombs, ready to take off for Russia on a secret signal from Washington. The sole aim of the modus vivendi was the greater defence and security of the United States.

But the Americans were shattered in August 1949 to learn that Russia had exploded a nuclear device. In principle, this now denied the US its hitherto unrivalled position as prime arbiter of international affairs. The news was also profoundly disturbing to the British government. The question was asked: 'How has Russia done it before us when they must have started from scratch and we had so much information from the outset?' The event blew away many illusions about the technological backwardness of the Soviet Union. It was obvious they had begun an atomic programme well before the end of World War II; it was a huge country and plants could have been built in the east when the Germans had been pushed back far enough to the west. This could well explain Stalin's cool and indifferent reception at Potsdam to the information about the Anglo-American bomb project.

The American government said the Soviets had been enabled to produce a bomb so quickly because of the transmission by spies of technical data from the Manhattan Project sites in the US. A witch hunt followed, led by Senator McCarthy, and this was of such severity that it terrified virtually all opposition into submission. Lesser waves of anti-communist fervour swept over Western Europe, especially following the outbreak of the Korean War in the wake of the final victory of the Chinese communist revolution in 1949. In February 1950, Klaus Fuchs, a senior scientist at Harwell who had worked at the heart of the Manhattan Project, was proved to have been a Soviet spy. Another blow fell for the British in September when Bruno Pontecorvo, another senior scientist, fled from Harwell to Russia. The Americans became very nervous about British nuclear security, even more so when Guy Burgess and Donald Maclean fled to Russia from the British Embassy at Washington in May 1951. But the Americans also had their own spies. The effect of the discovery of these spies was to cause the Americans to exert pressure on

Britain to introduce a system of 'positive vetting' of all nuclear staffs and potential recruits; in exchange, more nuclear information was to flow across the Atlantic. In practice, much of this information turned out to be somewhat dated in the experience of the present writer.

In February 1950 there was a general election in Britain. Labour was returned but with the very small majority of six. By now the Attlee administration was showing signs of lack of confidence and splits in the cabinet were emerging. However, at the end of 1950 the Chancellor was able to announce that Britain would no longer need to draw upon Marshall Aid; at the time it began in 1948 Europe was desperate for dollars. Public opinion spoke in glowing terms of the Americans' motive in coming to our economic aid. But the programme stemmed more from American enlightened self-interest than altruism. American producers needed the European consumers that Marshall Aid guaranteed and at the same time the US government decided that a prosperous Western Europe was ultimately the most effective barrier to the spread of communism. But Marshall Aid had not prevented another financial crisis erupting in 1949 and in September sterling was devalued by 30 per cent.

Although some said defence spending and overseas commitments were already too great for Britain's economy to bear, nevertheless Attlee's cabinet decided to increase spending dramatically to a total of £4,700 million in 1951-54. This was roughly 14 per cent of the total national income and it turned out to be far more than could possibly be spent. Some cabinet members thought that a war must be looming and welfare provision was severely cut. However, three members of the government could not agree to the latter; Aneurin Bevan, Harold Wilson and John Freeman all resigned.

The war led to a serious movement against Britain in the terms of trade and a steep rise in import costs which had deleterious effects on inflation. Britain's support for the United States in the Korean War was proving to be crippling economically and at one stage it looked as though the American government was on the point of using atomic weapons against North Korea and its Chinese allies. Attlee flew to the US in December 1950 to persuade President Truman against this course of action. Whatever passed between the two is not clear but fortunately the weapon was not used. MacArthur, the American general in Korea, was all for using the occasion to invade China herself in order to wipe out the new communist government; Truman gave him the sack! Finally, an armistice was agreed in 1953 and this divided Korea into a communist north and a non-communist south. In the meantime there had been a general election in Britain in October 1951, when Labour was defeated and Churchill had led the Conservatives back into power.

# Chapter 6

# The rush for Britain's nuclear bomb

When Churchill took over from Attlee in October 1951 he was obviously impressed with the nuclear project but he expressed a mixture of admiration, envy and shock that his predecessor had spent nearly £100 million on the atomic project without informing parliament (nearly a billion pounds in present money values). Churchill continued to operate in the same way without a qualm. He felt that Britain had been unfairly treated by the Americans and had been forced into spending such vast sums of money on her own project; he said Attlee had been 'feeble and incompetent' in not plugging the facts about the wartime nuclear agreements with the US. This was hardly fair, because Churchill's very tight security during the war had ensured that his deputy was kept unaware of all the nuclear negotiations.

Lord Cherwell was critical of Churchill on this matter, concerned as he was with the fact that it was now two years since the Russians had first exploded a nuclear bomb and it looked as though yet a further year would elapse before Britain would be able to do so, a gap of three years which was intolerable. He believed the slowness of the British programme was due to state control and the civil service machine and its bureaucracy. It was totally inadequate and he wanted an independent corporation. It is strange that it never seemed to have struck Cherwell that the Russian atomic programme would have been state-run and labouring under a bureaucratic machine. However, he got Churchill to go part way with him towards a degree of independence and the plan eventually emerged in April 1952 as a sort of hybrid affair which 'cut clean across the hallowed canons of ministerial responsibility', according to Margaret Gowing. Cherwell wanted a completely independent organisation which he achieved in 1954 when the United Kingdom Atomic Energy Authority (UKAEA) was established. G. R. Strauss on the Labour benches led strong opposition to the bill but in March 1954 the Conservative government won by 266 votes to 244.

## Operation Hurricane

In Chapter 4 we described how Attlee's atomic project developed inside three organisations: production under Hinton, atomic weapon development under Penney, research under Cockcroft. In January 1951 the pressure was on Hinton to produce fissionable material sufficient to allow Britain to explode an atomic bomb in the autumn of 1952. When this date was fixed by the Chiefs of Staff Hinton would have raised no objection for he loved to have to work to a deadline and thus to be able to drive his staff along against tight time schedules. Already he had become impatient over the research backing he received from Harwell and had therefore set up his own research and development (R & D) organisation, independent of Cockcroft.

The project for making the bomb was kept very secret indeed. It was coded Operation Hurricane. When the present author took up a post in January 1951 and commenced research on new methods for the preparation of fissionable materials, in the brand-new R & D laboratories at Windscale near to the newly-built nuclear piles, no mention was made of the intention to manufacture weapons. It was tacitly assumed that our objective would be power production in nuclear piles fuelled with plutonium.

By April, Pile No 1 was working on normal power but Pile No 2 was not commissioned until October. At the beginning of the following year, the first batch of irradiated uranium 'slugs' from Pile 1 were fed into a chemical separation plant where, after dissolution in nitric acid, the tiny amounts of plutonium present were separated. But that was not the end of the story. For the dilute plutonium solution was heavily contaminated with dangerous radioactive fission products and large quantities of uranium were still present, so the liquid, contained in appropriately-named lead coffins, had to be transferred to a further chemical plant for purification. And this was where things went wrong. The degree of purification required for bomb-grade plutonium was far greater than what was necessary for a civil power reactor to generate electricity. The nuclear bomb purity specification was very stringent and the plant could not achieve it. Furthermore, it seemed that some of the plutonium which the physicists had calculated should have been present, was not there. There was a crisis because it looked doubtful that the deadline for the explosion would be met owing to lack of the essential fissile material.

But the Prime Minister had issued a priority directive concerning Operation Hurricane and the works general manager at Windscale, one Henry G. Davey, normally a calm character, was beginning to feel desperate: he appealed to the R & D laboratories on site for help. As a result the present writer was put in charge of a task force of research chemists, analysts and engineers who took over responsibility for Building 209, which contained the plutonium purification plant. The site drawing office and engineering workshops along with the Chemical Inspectorate's analytical laboratories were to give our requirements top priority. The team worked night and day, on a plant that was very radioactive due to the batches already fed through it. Cherwell paid us a visit to boost our morale. We had to redesign the samplers in order to get correct information as to what was happening. We used data from secret American reports and from Harwell chemists. Eventually, we had 209 up and running in time for enough plutonium solution of adequate purity to be fed to the Finishing Plant to be turned into the solid state needed, in turn, for the Aldermaston plant to fabricate a nuclear bomb for detonation at Monte Bello by the chosen date.

## After Operation Hurricane

The special work carried out by R & D for Production Department in Building 209, and indeed in certain other plants, was fully recorded in official documents. However, Gowing gave me to understand that no information about the technical problems associated with Operation Hurricane was given to her by Hinton when she began to write up the official history of Britain's nuclear project. (One wonders just how reliable the official history of Britain's nuclear project can be if Hinton held back from it whatever he so desired.)

Shortly after Hurricane, I was called over to take tea with Davey. I politely congratulated him on his recent award of an OBE and he warmly thanked me for my part in Hurricane. Nobody in R & D had received a 'gong' but the manager of 209, who was on sick leave during the crisis, had received an MBE. Davey was well aware that I knew this but he told me he was a much older man and that I should not feel disgruntled for I was still relatively young and my time would come. But that was by the by, for what he had really called me over for was to help him draw up a Capital Expenditure Proposal (CEP) to cover the thousands of pounds expenditure incurred by the R & D Project Team during Hurricane. As a young man inexperienced in the ways of the Civil Service I could not see the point of wasting time putting up a proposal to a finance committee for something already done, and at the express wish of the PM. Davey laughed and said the National Audit would make a great fuss if we did not gain approval for any capital project we had decided to carry out, Prime Minister or no Prime Minister.

Operation Hurricane was only the beginning; it turned out that a whole series of test explosions were to be carried out by the British on the other side of the world. These were said to be safe and would not harm native peoples on nearby islands. But if they were that safe, why were they not carried out on home territory? In the event, people did suffer radiation as a result of the British series of tests, even including British servicemen in Australia, but getting the government to admit this turned out to be an extremely difficult task.

Exploding a nuclear device in October 1952 was of course a prestige exercise by Britain. But it did not mean that she was in any position to mount a nuclear attack on an enemy country or engage in nuclear war; she did not possess a nuclear deterrent. For that to be achieved, Britain would require a stockpile of bombs and a means of delivering them to their targets. In 1953, the RAF began to take delivery of nuclear weapons because it was assumed that the only practical means of delivery at that time was by aircraft. A new type of bomber was therefore required.

Although money and resources for the bomb development had been given the highest priority, this was not paralleled with regard to bomber development and in fact the RAF did not receive the means to deliver its new bombs until 1957. The first Valiant prototype had crashed in January 1952 and although the second prototype flew in April, it had been grounded by the time of the first Monte Bello explosion. The Valiant did not go into service until the end of 1954 and the first squadron was not operational until 1957, by which time more advanced types of V-bomber were beginning to become available. So there was a gap of about four years between Britain's first nuclear bomb test and the provision of any facility for using it. How did this gross imbalance arise? Clearly there was no need for such a crash programme for bombs, and Operation Hurricane was therefore quite unnecessary.

# Chapter 7

# The moment of hope ... and afterwards

After the Soviet Union ended America's monopoly of atomic weapons by exploding its own bomb in August 1949, there were strident demands in the US for a reply in kind; to keep 'ahead of the Russians'. There was a movement in favour of developing the H-bomb (see the end of this chapter for a brief technical note). Edward Teller of the Manhattan Project had already done sufficient theoretical work to suggest that it would be feasible to develop such a bomb. The General Advisory Committee of the US Atomic Energy Commission was asked to discuss the proposal and in October 1949 it came back with a report which rejected the idea. It was not considered necessary for national security, was of little use for military purposes and was morally wrong. Also, in answer to the argument that the Soviets might develop such a weapon, the committee replied, 'Our undertaking it will not prove a deterrent to them'. But the committee's advice was ignored, the chairman Robert Oppenheimer resigned and was indeed harassed by the witch hunters of the UnAmerican Activities Committee.

The H-bomb development project went ahead in America under Edward Teller's direction. When the so-called 'Mike Shot' was carried out on 1 November 1952, the force of the explosion was 10 million tonnes of TNT, roughly equivalent to 800 Hiroshima bombs. The implications of this development were then almost too horrible to contemplate. Although publicity in favour of the bomb said it would be a 'clean' weapon as it depended for its action on fusion rather than fission and there would be very little radioactivity associated with the explosion, this was false information.

If Mike Shot had been intended as a deterrent to the Soviets, it did not work. Nine months later, in 1953, they exploded in Siberia what appeared to be an H-bomb, according to the US radiation detection aircraft which revealed traces of lithium, an element present in certain types of H-bomb material. This in turn only served to spur on the Americans and on 1 March 1954 at Bikini Atoll, they carried out 'Bravo Shot', an explosion which was well over a thousand times that of the Hiroshima bomb, namely 14.8 mega-tonnes in force. (There was associated radioactive fall-out, the result of which is described in the following chapter.) There might have been doubts about the Russians' 1953 test being that of a true H-bomb — but they certainly had an operational H-bomb ready in 1955.

But it appeared that the horrific implications of these developments were becoming realised internationally and there was growing concern over them. Even the leaders of the superpowers seemed at last to be coming to their senses and were prepared to talk together once more.

## The moment of hope

Surprisingly, the initiative for controlling nuclear weapons internationally now came from the Soviet Union and it happened thus. Joseph Stalin had died in March 1953 and changes then took place in the Soviet Union which had far-reaching effects. His successor, Georgi Malenkov, was forced to resign and over the next two years, the new First Secretary, Nikita Khrushchev, worked to restore the Soviet Communist Party's role in decision-making and to establish its control over the whole state apparatus. Information was coming out about the scale of Stalin's repression of opponents and others and in 1955, after a special commission of the Party Central Committee had investigated the purges, Khrushchev roundly denounced Stalin and his regime. This was followed by radical changes in foreign policy as Stalin's isolationist policies were successively reversed.

May 1955 was a particularly significant month; although West Germany formally entered NATO on the 10th, the Americans put forward proposals for accepting force levels and also a draft programme for linked conventional/nuclear disarmament. Thus they were accepting for the first time the principle of an international control and verification agency. In response to this, the Soviets put forward the most detailed system yet advanced for its implementation, namely, a wide network of land-based UN observer posts. The Soviet Union also favoured an immediate ban on testing nuclear weapons. It began to look as though the entry of Britain into the nuclear arms league plus the production of the H-bomb in the USA and also in the Soviet Union might be having the effect of getting the leaders round the table to plan for nuclear disarmament in all seriousness. Lord Noel-Baker, the experienced British disarmament negotiator, called this 'the moment of hope'.

Also in May 1955, the Soviet Union signed a peace treaty with Austria and withdrew its troops from that country. Then Khrushchev flew to Belgrade and effected a rapprochement with Yugoslavia, also in May. But in the following month he responded to West Germany's entry to NATO by getting eight East European countries to sign a treaty of mutual assistance, thus setting up the Warsaw Pact.

In the August of 1955, world leaders met in Geneva in a spirit of hope following the Soviets' encouraging moves and the indications in May of the possibility of some form of East-West agreement on nuclear matters. But these hopes were dashed when Stassen, the US delegate, came forward with a new proposal on behalf of Eisenhower. It was for a hybrid plan of control which combined the US-favoured aerial surveillance with the Russian proposal of land-based control posts. It was not well received. The French sought to achieve a compromise but the US only hardened its position and in fact Stassen vetoed progress by placing a reservation on all of its pre-Geneva substantive positions. The moment of hope had passed and no definitive explanation of the line taken by the Americans at that summit has ever emerged. At no time since that summer of 1955 has the world been so near to realising general disarmament and international control of nuclear weapons. The Americans should surely have responded more warmly to the new Soviet foreign policy, which continued to develop so encouragingly even after the failure of the summit.

Straight after the Geneva meeting, the Soviets unilaterally reduced their armed forces by 640,000 men in recognition of a wish to 'recognise a lessening of international tension'. In September they removed their troops from Finland and also formally recognised West Germany. At the Party Congress in February 1956, the policy of 'peaceful co-existence' formally superseded the doctrine of inevitable war between capitalism and communism which had dominated Stalin's era. Sadly this had come too late because Western nations were hardening their policies. For example they were refusing to give an unqualified 'no first use' with regard to nuclear weapons, a policy to which the Soviets attached great importance. The Cold War would seem to be going to continue for a long time and indeed relations between West and East were to deteriorate still further in the following year.

## The sad events of 1956

It is important to recall the main sequence of events in 1956 which took international relations to a very low level.

First of all, John Foster Dulles, US Secretary of State, precipitated the 1956 Middle East crisis by unilaterally withdrawing his country's support for the Aswan High Dam project in Egypt on 19 July. Then Britain immediately withdrew her support for the project. President Abdel Nasser of Egypt responded on the 26th by nationalising the Suez Canal, pointing out that the High Dam could be built with its revenues. In the event, the Soviet Union provided Egypt with the necessary financial support. However, Nasser's action finally led to the ill-fated Anglo-French Suez invasion of 31 October to 6 November. There was of course much division in Britain about the Suez invasion. Later, it emerged that the action was passionately and privately opposed by Louis Mountbatten, the Chief of Staff instructed to carry out the operation. The American administration was also opposed to it but Dwight Eisenhower kept a low profile on the subject until after the presidential election.

Nevertheless it appears that the US engineered a run on the pound, leading to the British government going to the International Monetary Fund for a loan. The Americans had influence in that quarter and the loan was only agreed upon on the condition that Britain called an immediate cease-fire in Egypt. Harold Macmillan, Chancellor of the Exchequer, threatened to resign unless the IMF loan could be obtained; the cease-fire took place and thus the loan was secured.

Events taking place in Eastern Europe paralleled those in the Middle East. Mass demonstrations were taking place in Hungary on 22-23 October in support of a demand for the withdrawal of Soviet troops and the reinstatement of the former Prime Minister, Imre Nagy. After fighting took place in Budapest on 23-24 October a National Government was set up under Nagy and he requested the withdrawal of Soviet troops. The USSR said it would comply with this request, even though Hungary was a member of the Warsaw Pact. But then on 31 October, Nagy also declared that his country now wished to withdraw from the Pact and pursue an independent course. On the previous day in the House of Commons, Hugh Gaitskell, Leader of the Opposition, had demanded of Anthony Eden, the Prime Minister, upon what authority he was proceeding with his Suez operation. Gaitskell went on to add that nothing in the UN Charter justified any nation appointing itself as world policeman:

'... the great danger of the present situation is that if we can get away with this sort of action then so can anyone else.'

How right he was, for the hawks in Moscow, believing that their Warsaw Pact was in imminent danger of crumbling, were able to persuade Khrushchev to go into action in Hungary. Hence on 4 November, as the Suez war was in full spate, Soviet tanks and artillery attacked Budapest; Suez gave the Soviets both precedent and cover. However, the uprising was only put down with the loss of 50,000 Hungarian lives.

How low had international relations now sunk since that 'moment of hope' in the summer of 1955, when the world seemed to be so close to realising general disarmament. Where does responsibility lie for this terrible decline? Although it cannot be attributed to a single quarter, it can be seen that Britain must share the blame to some extent.

*Technical note*

The distinctive feature of the H-bomb is that nuclei of the heavy isotopes of hydrogen, namely deuterium (atomic weight 2) and tritium (atomic weight 3) become fused together rather than split as is the case with the simpler A-bomb or fission bomb. However, these heavy hydrogen atoms can only fuse at extremely high temperatures such as are reached in fission bombs. So a limited fission explosion of U-235 is used to trigger a fusion explosion by virtue of the heat generated. The 'thermo' part of 'thermonuclear', the alternative name for the H-bomb, refers to the importance of the high temperature. The fusion process involves a small decrease in total mass and the latter appears as an enormous amount of energy, about a thousand times the energy released in the explosion of the simpler A-bomb.

One way of packing together into a close space a large number of fusionable atoms is to employ the solid substance known as lithium deuteride. The detection by American aircraft of traces of lithium in the radioactive fallout from the Soviet bomb test in Siberia in August 1953 suggested that it might have been an H-bomb experiment.

# Chapter 8

# Growth in public concern over nuclear weapons

I referred in the previous chapter to Bravo Shot, the first American testing of an operational H-bomb, which took place on 1 March 1954 at Bikini Atoll. Although this was deemed a great success by the Americans because of its huge force, not everything went well with the firing. The wind blew radioactive debris eastwards over three of the Marshall Islands whose population knew nothing about radioactivity and its possible effects.

## The voyage of the *Lucky Dragon*

A small Japanese trawler, the *Fukuryu Maru* — the *Lucky Dragon* — was out fishing for tuna east of Bikini. Just before dawn on 1 March, a member of the crew was startled to see the western sky erupt with a whitish-yellow light which turned into a flaming orange ball. He soon had the others out on deck, because the sun rises in the east! Soon they realised it must have been an atomic explosion. When a light drizzle fell, they rubbed it between their fingers and some tasted it; it consisted of a fine sandy ash, not rain. By now they had hauled in their nets and were heading for home. By evening some of the sailors were listless, some were sick, and others had open sores and were losing hair. The crew of the *Lucky Dragon* finished up in hospital and two weeks after exposure they were found to be radioactive. Their illness and the death of one of the crew made headlines in Japan and strained relations with the US for quite some time.

The story of the *Lucky Dragon* brought the question of radioactive fall-out to the attention of the world's press and hence to popular attention probably for the first time; there had not been much interest in the media over the question of radioactive fall-out after Hiroshima, or only for a very short time, and the subject had certainly been neglected in the following nine years.

Samples of the fall-out from Bravo Shot collected from the deck of the *Lucky Dragon* were examined by nuclear scientists not involved in the American tests. As we have already noted, the H-bomb spreads radioactivity since it is basically a two-stage system. But in the Bravo Shot weapon there was even more to it than that. When Professor Joseph Rotblat, who had left the Manhattan Project in order to specialise in radiation medicine, heard of the finding of traces of an unexpected isotope in the Bravo Shot fall-out, he deduced correctly that the weapon had indeed been a three-stage bomb: fission, fusion, then fission again due to a coating of natural uranium. This third stage had added considerably to the amount of dangerous radioactivity in the bomb's fall-out.

## Parliament and the H-bomb

Following Bravo Shot there was a short debate on the thermonuclear bomb in the House of Commons, on 5 April. There was concern about the international tension and Labour called for a summit of nuclear powers to try and defuse matters. This was accepted by Churchill but nothing came of it. The H-bomb did come up for debate again in the House in the following year and Churchill supported it on the grounds of deterrence. Attlee, still Leader of the Opposition, agreed with him. As ever on nuclear matters, Labour and Conservatives were united in the Commons, but some dissent did emerge on this occasion. Aneurin Bevan from the Labour benches, whilst being careful not to endorse a clear unilateralist position, intervened to criticise Attlee's lack of leadership over the matter and received the support of 62 MPs. On 16 March 1955, the Parliamentary Labour Party decided to nip this in the bud and by 141 votes to 112 actually agreed to withdraw the whip from Bevan. The Labour right was really determined to get rid of him and the case then went to Labour's National Executive Committee (NEC), but there it was agreed by 14 votes to 13 not to expel Bevan from the Labour Party. But this was not before Attlee had reportedly received assurances from Bevan with regard to his future conduct. On 30 March, the Labour NEC called on the government to develop a British thermonuclear weapon 'to consolidate Britain's independence.' It was quite possible that work on developing an H-bomb was already secretly under way. On 28 April, Bevan was re-admitted into the Parliamentary Labour Party.

## Nuclear strategy

On 15 May 1957, Britain announced that it had exploded a thermonuclear bomb. So we now had Britain, the United States and the Soviet Union equipped not only with nuclear fission bombs but also with thermonuclear bombs. They all claimed that it was for reasons of deterring an aggressor and not because they would ever be the first to use these weapons. The reasons for adding thermonuclear weapons to their 'conventional' stock was because they had to keep up with a potential enemy. Thus the arms race continued; it had similarities with the arms race between Britain and Germany prior to World War 1 when each side continued to build more and more battleships — and better ones, too!

But a world war which started non-nuclear would be unlikely to end without becoming nuclear; one side would use nuclear weapons either to avoid defeat or to accelerate the achievement of its victory — as was the logic for the American bombing of Japan in 1945. But the military strategists in the USA now estimated that about 70 per cent of a Russian attacking bomber force would be able to penetrate America's air defences and so one of the basic concepts of the nuclear deterrent went out of the window; America's industrial strength was concentrated in known locations largely round its geographical fringe while Russian industry was dispersed throughout Asia. So the Americans started to look for an alternative to an all-out nuclear exchange. The idea emerged that the Western allies would deliberately aim to restrict a war with the Soviets to attacking military targets in the theatre of land-fighting, otherwise it would be difficult to prevent a rapid transition to the indiscriminate use of thermonuclear weapons against enemy cities on both sides. This now led to the development of smaller nuclear weapons, to be fired from guns on the battlefield. This was to be termed the strategy of 'graduated deterrence', later known as 'flexible response'; and it was eventually to become the adopted war-fighting strategy of the NATO allies. It hardly needs saying that mainland countries in Europe, notably Germany, were not happy with the idea that their territories would become radioactive wastelands in the event of war between west and east. Also, exercises in Germany began to point to the probable deaths of millions of civilians in that country. Nevertheless the production of battlefield nuclear weapons went ahead regardless, Britain included.

Setting aside for the moment the question of continued American and Soviet development and stockpiling of nuclear fission and H-bombs, there was really no case for Britain to do so. Clearly Britain was indefensible in the case of an attack by a nuclear aggressor and to threaten to use thermonuclear weapons as a first-strike or a retaliatory strike against a country holding such weapons would have been suicidal. The use of thermonuclear weapons against a country without any nuclear weapons would be sheer genocide. This must surely have been realised within the inner circles of government and Chiefs of Staff and so the only reason for going on with nuclear weapon development and production was mere national prestige; the very momentum of work at Aldermaston and its associated weapon stations would have had a power which government quailed at the thought of terminating or even running down. Not only did the Conservative government and the Labour opposition continue with their bipartisan policy on nuclear affairs, but clearly the Labour leaders were determined to stamp out any sign of resistance from the antinuclear lobby within the party. But opposition did develop within the Labour movement and even the trade unions, as we shall see. And there was clearly a rising antinuclear feeling outside party politics in Britain in the late 1950s.

Since 1953 the levels of radioactivity in the world's atmosphere had been increasing due to nuclear weapon testing; this information must have worried the government and the fact was therefore kept secret. To try and learn more about the effects of radioactivity on the human body, it seems that secret experiments of a macabre kind were carried out on people without their knowledge. If this had been known to the public there could well have been much greater feeling against continuing with the weapons programme. Even greater opposition would inevitably have developed if the public had been aware of the accidents involving nuclear bombs in England in the late 1950s. But these events were successfully kept secret by the government for nearly forty years; we describe and discuss what is now known of them in the next chapter.

Nuclear piles were also possible sources of accident. In 1957, No 1 pile caught fire at Windscale and contaminated a rural area with radioactive fallout. This incident became widely known, mainly because many gallons of milk in Cumbria became contaminated with radioactivity and had to be wasted. However, it was to be thirty years before the government allowed some (but not all) the details of the accident to be made public; in Chapter 10 we tell the story of the fire and its sequence.

## Unilateral nuclear disarmament

During 1957, there was a growing distinction between those on the one hand who supported 'multilateral nuclear disarmament' and on the other hand those who desired 'unilateral nuclear disarmament'. The basic idea behind the multilateral view was utopian inasmuch as it depended on all states with nuclear weapons agreeing jointly to give up their weapons. But by now it was clear that this was not going to happen in practice. Supporters of this view privately knew this was so but the title of the policy implied that it would be worked for and eventually come about. In the meantime, it was business as before and everyone kept their weapons and relied on the deterrent strategy. But this was a dangerous strategy and meant that people had to continue living under the ever-present threat of war; the Cold War might become a Hot War, and nuclear catastrophe could result.

The unilateralists were no longer prepared to tolerate this policy; they believed their view was not utopian, rather was it seen by them as a positive, practical strategy which could begin to operate by one country deciding to give up its present stance by steadily abolishing its nuclear weapons capability. The view that other countries would immediately follow suit might be seen as utopian but to some people it seemed the only thing to try; it was perhaps a policy based on desperation. If the policy were to be adopted, however, it was unlikely that a move in that direction would come from the British government or even from the opposition in parliament; it would therefore need to develop by extra-parliamentary means. Articles arguing the case for unilateralism appeared in the press and indeed it became the editorial policy of the *Daily Herald*, at that time the daily paper of the Labour left and the trade unions. Alarm bells began to ring at Labour Party HQ and the constituencies were 'warned' against the spread of unilateralism. At the Labour Party conference in October, Aneurin Bevan caused consternation amongst his traditional left-wing supporters by successfully opposing a unilateralist motion renouncing nuclear weapons. However, in the previous month the Liberal Party Assembly had called for Britain to unilaterally abandon the H-bomb.

### The Campaign for Nuclear Disarmament

Many people not directly involved with party politics in Britain were concerned about the situation and already there was in existence an organisation which was strongly opposed to any further testing of nuclear weapons — the National Committee Against Nuclear Weapon Testing (NCANWT). In 1957 the Direct Action Committee against Nuclear War (DAC) was formed. But such groups were soon to be overtaken by the establishment of the Campaign for Nuclear Disarmament (CND).

It is interesting to note that many eminent people were associated with the setting up of CND. In November 1957, a small group met at the home of Kingsley Martin, the then editor of the *New Statesman*. Among those present were the author J. B. Priestley, his archaeologist wife Jacquetta Hawkes, Prof. P. M. S. Blackett, the leading physicist who twelve years earlier had exerted pressure on Prime Minister Attlee in a vain attempt to persuade him against starting a nuclear weapon programme, as well as Bertrand Russell. They discussed the virtue of a major campaign for the abolition of nuclear weapons. This was followed up by a meeting on 15 January 1958, attended by some 50 distinguished people, including Sir Julian Huxley and Dame Rose Macaulay. The outcome was the formation of CND with Canon Collins as its first chairman. NCANWT now yielded the primacy gracefully and transferred its funds to CND. The first executive of CND included Ritchie Calder (vice-chairman), James Cameron, Arthur Goss, Kingsley Martin, J. B. Priestley, Prof. J. Rotblat, Sheila Jones and Peggy Duff (organising secretary). Co-opted later were Sir Richard Acland, Frank Beswick, Benn Levy and A. J. P. Taylor. Most were aged between 45 and 65, hardly an outfit of the young! Several had been publicly identified with the Labour Party, chiefly as rebels. Lord Russell became its president. CND tried to keep open lines of communication with the Direct Action Committee. The Campaign also possessed 38 sponsors at the early stage; again many were well known and included Lord Boyd Orr, John Arlott, Benjamin Britten, Dame Edith Evans, Rev. Trevor Huddleston, Sir Compton Mackenzie, Henry Moore, Flora Robson, and Barbara Wootton.

The inaugural meeting of CND was held in the Central Hall, Westminster and four adjoining overflow halls on 17 February, and it was decided formally on this occasion that CND would campaign for unilateral nuclear disarmament. Later a policy for CND was agreed and issued;

'We seek to persuade the British people that Britain must:

(a) Renounce unconditionally the use or production of nuclear weapons and refuse to allow their use by others in her defence.

(b) Use her utmost endeavour to bring about negotiations to end the arms race.

(c) Invite the co-operation of other nations, particularly non-nuclear powers, in her renunciation of nuclear weapons.

Realising the need for action on particular issues, pending success in major objectives, Britain must;

(a) Halt the patrol flights of planes carrying nuclear weapons.

(b) Make no further tests of nuclear weapons.

(c) Not proceed with the agreement for the establishment of missile bases on her territory.

(d) Refuse to provide nuclear weapons for any other country.'

So far as immediate positive action went, the first activity was to take up the idea of a march to Aldermaston at Easter. DAC activists had originated the idea, so this first march was a joint effort. Although the weather was bad, 4,000 people took part. In the summer of 1958, CND held a mass lobby of parliament which about 9,000 attended. In the following year, the Easter march began at Aldermaston and ended in Trafalgar Square where 20,000 people were present. Support for CND rose rapidly and in April 1960 there were 100,000 supporters in Trafalgar Square.

But some were becoming impatient over the lack of response by parliament and government to pressure for action on CND policy and the 'Committee of 100' was formed at the instigation of Russell. This was a unilateralist group which committed itself to non-violent direct action. In September 1961 they organised a mass sit-down in Trafalgar Square, ending with the arrest of 1,300 people. Some members of CND did not approve of Russell's committee, believing that they should concentrate effort on winning Labour over to the anti-nuclear policy. In fact the Labour Party conference at Scarborough in October 1960 accepted the unilateralist motion by 3,300,000 votes to 2,900,000. Frank Cousins, a unilateralist, commented that 'we are the real patriots', but Hugh Gaitskill promised to get the decision reversed — he would 'fight, fight and fight again to save the Labour Party'. Indeed, at the Labour conference in 1961, he was successful in getting the decision reversed, by 4,000,000 votes to 1,700,000. Large swings in votes on unilateralism reflected the volatility of views in various trade unions, with consequent impact on block votes.

The high point of CND activity and support in the 1960s was the 1963 Aldermaston march, with over 100,000 gathering in Trafalgar Square, for there was a falling off of support after Lord Home signed the Nuclear Test Ban Treaty in July 1963 with the US and the USSR, to ban all 'overground' nuclear testing. It is noteworthy that this did not exclude testing below the earth's surface. But strong support for CND grew again some years later following events which are described in Chapter 24.

## The origin of 'Pugwash'

About the time when CND was being formed, Lord Russell was stirring the scientists into action. The Americans had exploded a thermonuclear weapon in 1954, as we described above, and British and Soviet scientists were also developing such a weapon. Russell wrote to Einstein to the effect that 'the moment had truly come when eminent men of science should draw the attention of world leaders to the impending destruction of the human race'. This led to the formulation and signature of the so-called 'Russell-Einstein Manifesto' coupled with Russell's wish that this should be formally launched at a major public meeting.

At this stage, Russell had got to know Professor Rotblat who was then vice-president of what was called the British Atomic Scientists' Association, and he asked him to chair the meeting at Caxton Hall, which was very successful and drew a large crowd. It was out of this gathering that the concept arose of a conference of scientists to assess the precise nature of the danger from nuclear weapons and also to try to involve the international community. One Cyrus Eaton, President of the Chesapeake and Ohio Railroad, now offered to finance the proposed conference at his home in Pugwash, Nova Scotia.

However, Eaton stipulated that the name 'Pugwash' be used in the conference title; hence the inevitable confusion with the much better known Captain of that ilk!

The first conference duly took place in July 1957 at Pugwash and was attended by twenty one scientists, mostly physicists, from 10 countries. Rotblat was appointed secretary-general and the public announcement at the conclusion of the three-day meeting ended with the words:

'We are convinced that mankind must abolish war or suffer catastrophe.'

Since 1957, there have been some two hundred general and special meetings in the international series of 'Pugwash Conferences on Science and World Affairs', held in all parts of the world and with participants from sixty or more countries, all attending in a personal capacity. Pugwash deliberately shunned high-profile public campaigning but it was influential in the early days in achieving agreement on the 1963 Partial Test Ban Treaty (see Chapter 21) and more recently in scientific aspects of the negotiations leading to the ending of the Cold War. The 1995 Nobel Peace Prize was awarded jointly to the Pugwash Conference and to its president, Professor Joseph Rotblat. The British Pugwash Group has continued to be strong since its inception 40 years ago, largely because of Rotblat's influence in it.

# Chapter 9

# Accidents with nuclear weapons

Radioactivity levels were increasing in the earth's atmosphere during the 1950s but as we saw in Chapter 8, attempts by government to keep this secret were not wholly successful and public concern over the matter led to protest. Although it was generally believed, in Britain at any rate, that the steady increase in radioactivity was due to the testing of nuclear weapons in the atmosphere, it was also caused by various nuclear accidents, the details of which were kept secret for very many years. Undoubtedly the most serious of these occurred in the Soviet Union in 1957 when a storage tank in a nuclear processing plant near Chelyabinsk in the Southern Urals exploded and thus contaminated an area approximately 100 kilometres long by 10 wide. This led to the evacuation of over 10,000 inhabitants whose villages were destroyed. The British government claimed not to know of this accident for many years although it is probable that the UKAEA knew about it at an early date. This incident would have added to the level of radioactivity in the atmosphere over Britain. But by coincidence a serious fire occurred inside a British nuclear pile at the Windscale site in the same year as the Chelyabinsk explosion. Much information became known about the Windscale fire many years later, although it had been hushed up in 1957, and our next chapter is devoted to it and its consequences.

The present chapter is largely concerned with accidents relating to nuclear bombs at airfields in Britain. The public generally knew next to nothing about these until comparatively recently because the Ministry of Defence has always had a policy, jointly with the Americans, of maintaining the utmost secrecy about such incidents. Indeed they still have, unless circumstances reveal facts which force an admission from government. It was only in the 1990s that information began leaking into the public domain about certain accidents, notably at Greenham Common air base. It is only possible to describe some of the airfield incidents here; we cannot be sure there were no others and indeed the Campaign for Nuclear Disarmament believes there have been at least twenty in Britain. The Pentagon admitted to a total of thirty-three major accidents involving US nuclear weapons by 1968; this would also have included those in Britain.

Indeed, there were more US nuclear weapons than British-made ones on our military airfields in the 1950s and early 1960s. After the Korean War ended in 1953, the United States Air Force (USAF) was greatly expanded. In particular, its Strategic Air Command (SAC) was built up into a powerful nuclear striking force equipped initially with B-47s, armed first with atomic and later with thermonuclear weapons. Since these aircraft had insufficient range to reach the USSR from the American mainland a ring of forward air bases was established in such countries as Alaska, Great Britain, Spain, Turkey and Japan from which any part of the Soviet Union was accessible. Actually, US strategic bombers had been given permission to be stationed at airfields in Britain by the Attlee government as far back as 1948 (see Chapter 5) It is interesting to note what Winston Churchill said in 1951 about Attlee's decision:

> 'We must not forget that by creating the American atomic base in East Anglia, we have made ourselves the target and perhaps the bull's-eye of any Soviet attack.'

## Nuclear bomber accidents at British airfields, 1956-1963

I have chosen this period because it is precisely the time during which CND was conceived and grew in size, culminating in the Aldermaston March of 1963 when 100,000 people gathered in Trafalgar Square to protest against Britain continuing to keep weapons. If the facts of these accidents had been made public at the time, it is a fair bet that support for Britain banning nuclear weapons would have become overwhelming and the course of this country's history might have been different...

On 27 July 1956 at RAF Lakenheath in Suffolk, a B-47 American nuclear bomber crashed into a store containing three nuclear weapons. The plane's fuel caught fire but fortunately the TNT (conventional high explosive) contained in the triggers of the nuclear weapons failed to ignite. It was not until 1979 that a report on the crash leaked out — in the *Omaha World Herald* in the US — but the military denied it. In Britain, the Defence Ministry merely commented that the crash was contained by the emergency services without hazard to the civilian population. (What about the station personnel?) Writing about the accident in America, a retired USAF general who was in Britain at the time expressed the view that if the triggers had ignited, 'part of eastern England would have been turned into a (nuclear) desert.' Thousands of pounds of exploding TNT would certainly have scattered dangerous radioactive material from the bombs over a very wide area. At the time of writing the present book, an unabridged and honest report on the accident from the Ministry of Defence is still awaited.

In August of the following year, a rather similar incident occurred at Newbury in Berkshire after a B-47 bomber stationed at Greenham Common suffered engine trouble and jettisoned its wing fuel tanks onto the base. A serious fire resulted. No official record of the accident has yet been published but a survey of the old base in 1994 revealed signs of there having been a fire in a hangar. Local people knew something serious had happened because they saw signs of a fire and then found the base was closed for the rest of the month because of a series of 'emergency measures'. But the USAF released no pertinent information. The local paper, the *Newbury Weekly News*, reported that a public open day at the base had to be cancelled. The US base commander made a statement to the newspaper to the effect that there had been an emergency following a massive spill of 2,000 gallons of fuel spirit near to six aircraft; he said nothing about a fire. But even that information does not appear to have become widespread, Macmillan had made his 'you've never had it so good' speech in July and the British had therefore gone away on holiday in August with this ringing in their ears. Nothing more was revealed about the incident for 39 years, when a leaked classified document indicated that high levels of radioactivity due to highly enriched uranium had been found around the base when scientists from Aldermaston had carried out a survey of the area in 1960.

In January 1958, the *Daily Telegraph* reported that the American government had confirmed that a nuclear bomber had crashed in the US, spreading radioactive contamination during the course of the ensuing fire. Then in the following month, on 28 February, a serious fire appeared to have taken place at the Greenham Common base, again involving a nuclear bomber. The *Newbury Weekly News* showed a photograph of a burnt-out bomber at the base, taken by an amateur photographer who probably narrowly missed being arrested. A minor comment on the incident was obtained from the government at the time, strenuously denying that any nuclear bombs were involved.

Then in May 1959, incredibly, a 2,000 lb nuclear weapon fell from a Vickers Valiant bomber based at RAF Wittering in Cambridgeshire. Nothing was revealed about this incident for 37 years, until in August 1996 the 49 Squadron record book at Wittering for May 1959 was declassified and lodged at the Public Records Office in Kew. That record book refers to Exercise Mayflight and states that 'a 2000 lb nuclear weapon was accidentally jettisoned and severe damage resulted to the weapon'. An MOD spokesperson commenting in 1996 said it would have been an inert dummy as used for training, adding that the wording of the station commander could not be explained now because it was so long after the event. This statement revealed that even after all these years, the MOD still does not intend to give away any more information about nuclear weapon mishaps than it is forced to.

Returning to the matter of the nuclear weapon incidents of 1957 and 1958 at the USAF airbase at Greenham Common, the Americans had remained tight-lipped about them, even to the extent of failing to inform their neighbours six miles to the East — at the Atomic Weapons Research Establishment (AWRE), Aldermaston. Indeed, as we shall see in Chapter 12, until after Congress ratified a change to the US Atomic Energy Act in June 1958 there was very little contact between the US and British on matters including nuclear weapon materials and design. But the chemists at AWRE were to find out for themselves, by accident, some very important information relating to at least one of the fires at Greenham involving a nuclear weapon.

It was quite obviously an irresponsible plan in the first place to operate nuclear bombers from an airfield so close to where nuclear materials were collected and warheads prepared. So when the AWRE Aldermaston chemists identified radioactive contamination caused by highly-enriched uranium-235, a fissile material found in nuclear weapons, outside their perimeter fence in 1959, they would have been very suspicious of where it had come from. Further sampling soon traced it to the vicinity of Greenham Common where 20 grams of the U-235 had been deposited in a giant butterfly pattern with the airbase at the centre of the body. The level of enriched uranium was well above what might have been expected from nuclear weapons testing in the earth's atmosphere. Plutonium-239, another constituent of bombs, was also tested for, but the levels were not above what would have been expected from aerial weapons testing.

The AWRE chemists recorded their results in a report issued in 1961 and a copy of this was sent to Sir William Penney, formerly director of AWRE, but by then the UKAEA's Board Member for Research and Development. The report was of interest to him because it contained information which would reveal to an expert, data about the Americans' latest nuclear weapons. Further sampling of an even wider area than Newbury revealed that the element lithium, a constituent of some thermonuclear bombs, had been widely scattered. AWRE drew the conclusion that contaminated dust from at least one of the suspected fires had been spread from the runways as planes took off. This further explained the butterfly pattern of radioactive contamination. Three decades later, the MOD still refused to publish an official explanation of the data obtained by the Aldermaston chemists — perhaps the USAF never told them.

Accidents continued to occur. In 1961 a US bomber at an unidentified USAF base in Britain caught fire either during engine run-up or actual take-off. The plane, a B-47, was believed to be on 'airborne alert' when it would normally be carrying a nuclear weapon on the centre line pylon. There have been unofficial reports of the weapon being scorched and blistered and having to be replaced.

## Accidents in which nuclear bombs fell from USAF planes in flight

The present chapter is mainly concerned with accidents which had involved nuclear bombers at air bases in Britain. But only in the case of the Greenham Common base has it been proved, and by official papers released in 1996, that radioactive material from nuclear weapons has escaped outside the perimeter fence of the offending base. In 1961, a B-52 on airborne alert over North Carolina suffered structural damage and broke up, thus dropping two nuclear bombs. According to an 'insider', many of the safety devices on these weapons failed although officially the plane was said not to be armed. This was of course the official response one would have expected from the British government if a USAF bomber had crashed over here. But it raises the important point that if the insider was correct, then it was only by a whisker that America had escaped having a full nuclear explosion on her own soil. If that had happened, it could have been mistaken for a nuclear attack by the Soviets and would probably have resulted in immediate nuclear counter attack.

The USAF experienced at least two nuclear weapon accidents which involved widespread contamination of the environment and which it proved impossible to keep secret although no actual nuclear explosion took place. Neither of these incidents actually occurred on US territory. One involved a mid-air collision of a B-52, carrying several nuclear weapons, with a tanker aircraft during refuelling in January 1966. The planes crashed near Palomares in Spain. The conventional detonating devices of two bombs exploded, thus scattering plutonium over the landscape beneath. About 1,000 tons of radioactive soil and vegetation had to be removed. A third bomb fell into the Mediterranean and was later recovered from a depth of over 2,000 feet after extensive operations. The other accident, again involving the crash of a B-52, occurred when it was on airborne alert with nuclear weapons on board and flying near Thule, Greenland in January 1968. This was an area from which Denmark had banned nuclear weapons, but during the days of the Cold War, the super powers sometimes did not worry too much about the rights of small nations if they thought they could get away with it. The four bombs were destroyed by fire and the ensuing radioactive contamination resulted in nearly 250,000 cubic feet of material having to be removed. But the Americans could no longer continue to suffer the international disquiet about such incidents and at last Secretary of Defense McNamara forbade the carrying of nuclear weapons by planes on airborne alert. One wonders if this instruction was always carried out, especially at times of international tension.

## The legacy?

Following the Thule accident, 800 Danes were involved in the clean-up operation. Twenty years later it was reported that 500 of these had fallen ill, of whom 100 had suffered various types of cancer.

There has been no follow-up in Britain of the health of people who must have worked on the clean-up of the Greenham Common base after the fires in 1957 and 1958, let alone that of the inhabitants of those areas outside the perimeter which, in 1961, were found to be contaminated with uranium-235. But rumours of unusually high rates of radiation-linked cancers found around the nuclear weapon establishments of Aldermaston and Burghfield, only a few kilometres east of Greenham, began to circulate following letters to *The Lancet* medical journal in 1985 and 1986. These mainly referred to cases of acute leukaemia in children. This suggested a link with emissions from the nuclear establishments and in 1989 the government appointed its Committee on Medical Aspects of Radiation in the Environment (COMARE) to investigate these claims. No link was found. Greenham Common was not included in the COMARE study even though the government knew of the spread of radiation from the 1961 report, nor was COMARE's research team informed about the fires at the airfield. This was a pity; it might have proved useful.

When the 1961 reports on the AWRE tests around the Greenham area were declassified in 1996, many people in Britain were already familiar with the 'leukaemia clusters' located near certain nuclear establishments, the possible link with emissions being much discussed by the media. It was not surprising therefore that cases of leukaemia in and around Newbury just west of Greenham began to be heard of. Local families stated that there had been ten cases of leukaemia in the town near to the base, six having occurred since 1990; eight were said to be within 200 yards of the base perimeter. Following a couple of very crowded meetings held by the local council, it was decided to commission an independent survey by a university to check if there were still, nearly 40 years on, any unduly high levels of radiation on or around the base. A preliminary investigation did not detect anything untoward. But COMARE was to look at the childhood leukaemia situation in the area, after being given full information from earlier government reports.

# Chapter 10

# The Windscale pile fire of 1957

The Windscale fire of October 1957 resulted in the permanent closure of the original two air-cooled piles, Britain's sole sources of weapon-grade plutonium prior to the commissioning of Calder Hall in 1956. A report was drawn up by Sir William Penney. Only an abridged version (Cmnd 302) was published, by order of the Prime Minister, Harold Macmillan, who declared:

'It would provide ammunition to those in the United States who would oppose any amendment of the 1946 McMahon Act to allow more collaboration between the US and UK in the military application of atomic energy.'

They might argue that Britain was not to be trusted with more information about nuclear technology. But neither did the government wish to disclose to the Americans details of the military plutonium it was producing at Windscale.

Thirty years after the Windscale fire, much of the story was revealed in a set of secret papers released at the Public Records Office. Recognised as the most serious nuclear reactor accident to have occurred in Britain to date, clearly the fire had been more serious than had been admitted in 1957, inasmuch as the radioactive contamination was more widespread, even reaching as far as other countries in Europe. Writing in *The Guardian* in January 1988, Richard Norton-Taylor said it added up to 'a chilling picture of management incompetence.' Regrettably the Thatcher administration decreed that the papers recording the evidence of individual witnesses to the committee of inquiry should be held back for a further twenty years. Even so, there is enough information available to support Norton-Taylor's view of the incident.

## The sequence of events

Piles 1 and 2 at Windscale consisted of graphite blocks moderating the reaction of aluminium-encased uranium rods that could be pushed in and out along horizontal channels. During months of operation, irradiation of the graphite caused a rupture of bonds between carbon atoms and also strains in the lattice structure. This resulted in an increase of stored energy within the lattice. Under certain conditions enough stored energy could accumulate such that if released suddenly a considerable rise in temperature would take place. But the pile operators had to ensure that this never happened and, curiously, they did this routinely at intervals by deliberately allowing the temperature of the graphite blocks to rise just sufficiently to release the accumulated energy. Clearly this procedure, known as 'Wigner release', would be a tricky business.

No 1 pile was undergoing such an operation early in the October of 1957. The heating-up procedure had been duly carried out but had not proved successful, at any rate up to the morning of the 8th, so the operators decided to try and boost release by raising the temperature still further. Because of incorrectly sited temperature gauges the operators did not at that time realise that their actions were resulting in the overheating of the uranium fuel rods, which in turn led to ignition of the uranium and finally the surrounding graphite blocks.

In the early afternoon of the 10th, a reading of the radioactivity level in the chimney stack of No 1 pile showed it to be well above normal and further-more, the pile temperature was rising rapidly. At this stage, the factory management was notified. By early evening there was an obvious glow from the chimney. During that night, repeated attempts were made to create some form of fire-break by pushing out the horizontal fuel channels by applying brute force with steel poles. But the glow from the chimney got brighter and eventually flames emerged. The graphite had reached 1,000 deg C. By early morning they were running out of poles and therefore commandeered scaffolding poles from the nearby Calder Hall construction site. (Calder Hall nuclear power station is described in Chapter 11). During the early hours of the 11th, people on the Windscale site were warned of the emergency and told to stay indoors and wear masks. There was high airborne contamination with radioactive strontium and iodine. Work on Calder Hall was stopped.

It was decided to use water cooling in a desperate attempt to quench the now fiercely burning pile, despite the possibility of a hydrogen explosion. It was indeed a high risk strategy. At one point the pumping rate reached 1,000 gallons a minute. The Chief Constable of Cumberland was now contacted and warned of a possible emergency at the Windscale site. By the afternoon of the 12th it was concluded that the fire was out and all water flow was stopped. Both air-cooled piles were shut down, were to remain shut down, and would eventually be filled with concrete.

Samples of milk from the surrounding area were not collected until the

11th, the day after the fire started. Furthermore the results of the analyses were not available until the 12th, when the Ministry of Agriculture stopped distribution of local milk as from that evening. Initially the restriction was limited to a strip seven miles by two, but later this was extended to an area of over 200 square miles and for a period of six weeks. In response to complaints about the delay in getting the milk analysed, the official reply was: 'The laboratory in which this work is (normally) carried out was out of action because of the site emergency.' This was a feeble excuse. The health physics staff should have hastened to obtain samples of grass and milk and at the same time arranged to have a suitable laboratory standing by to do the analysis immediately on receipt of the samples.

It was because of the presence of the two radioactive isotopes strontium 90 and iodine 131 in the stack emission that speed was required. They would become concentrated in any cows eating contaminated grass and passed on to humans drinking their milk; the strontium would then mimic calcium and locate in bone with a risk of leukaemia and the iodine would collect in the thyroid gland with a chance of causing cancer there. Other radioactive isotopes were also found to be present in the atmosphere and on the ground and would augment the background level of radiation, thus increasing the risk of genetic damage within cells.

## The errors of management

From a White Paper on the accident published on 8 November 1957, plus the hitherto secret papers released towards the end of 1987 which included the unabridged version of the official report drawn up by Penney, one can extract the main conclusions about the Windscale fire and note the errors of management.

Some deficiencies originated from the engineering. The Windscale piles were designed and built in a hurry. It was only as a result of pressure from Sir John Cockcroft, director of the Harwell establishment, that the filters sited near the top of the pile chimneys had been installed. If the engineers had failed to do this, then the result of the Windscale fire would have been even more serious than it actually was. Inadequate consideration was given to instrumentation, in particular the siting of the temperature gauges. Other deficiencies stemmed from ill-defined operating procedures. It was clear that inadequate consideration had been given to the necessary operation of the Wigner energy release. They knew something about the phenomenon but a cavalier attitude prevailed and the operators of the pile did not have available a basic instruction manual. The people controlling the operation of the piles failed to alert their seniors at the first indication of things not proceeding normally with the energy release operation. Although the blame might have been placed on either the operators or the management it seems that the latter must accept most of the blame for not instructing staff in the correct procedures to take in the event of an unusual occurrence; this should always include picking up the telephone and contacting the next senior officer at the outset.

But one of the most serious criticisms levelled at the Windscale management is their failure to inform the local population at the same time that they told the workforce on site to stay indoors. They should also have been advised to exclude fall-out from their houses by closing all windows and blocking chimneys, this as a precaution until further advice could be given. Presumably the management had quailed at the risk of causing panic among the general public. But this risk has to be taken in such a situation. Indeed, it turned out that some 20,000 curies of the radioactive iodine 131 isotope were released from the pile. Although not disclosed until 1983 for reasons of national security, polonium was also released; it was being 'processed' as part of the H-bomb programme. It is highly dangerous even in minute concentrations.

Lord Penney's report, referring to the health implications, stated: 'It appears to us unsatisfactory that tolerance levels in respect of several of the possible hazards should have had to be worked out in haste after the accident had happened.' It had been claimed by the Medical Research Council, in an annex attached to the White Paper, that it was unlikely that any harm had been done to the health of anybody, whether a worker at the Windscale site or a member of the general public. But it was recognised that no work had been done on calculating the effects of sudden short-term exposure to radioactive fall-out. In fact the National Radiological Protection Board (NRPB), set up indirectly as a result of the Windscale fire, estimated that the accident could have caused up to 260 cases of thyroid cancer, 13 of them being fatal. The NRPB gave its figures on the possible consequences to health in 1983 when John Dunster, then the board's chairman, said that far from blowing out to sea as had been reported at the time of the fire, the main radioactive cloud travelled south-east over England.

## The Fleck Committee

Following the Windscale accident, there was a concern within the UKAEA that perhaps its management structure needed improving. Was management adequate for preventing such catastrophes occurring in the future and if they did occur, was the set-up adequate for taking quick and effective action to mitigate the effects? The UKAEA had only been in existence for three years. Prior to that, all nuclear work in Britain was under government control and as we saw in Chapter 6, Cherwell was unhappy about this and had exerted pressure on Churchill to create a completely independent organisation. Cherwell duly got his wish granted when the UKAEA came into being under the provision of the Atomic Energy Authority Act in August 1954. Now, after only three years, there were doubts about whether they had got it right. The UKAEA was independent of the civil service structure and was responsible for its own organisation and management; but it was not privatised, for it was financed by an annual grant from government.

The UKAEA set up a committee to examine its organisation and in charge of this was Sir Alexander Fleck who had been chairman of the chemicals group known as ICI. That company had long been proud of its structure and its quality of management. Fleck was an active and colourful personality with his own way of going about things. He was generally regarded as an ideal choice for the job but the staff of the various establishments and factories of the UKAEA were somewhat apprehensive about what he would do.

At Harwell, Cockcroft decided off his own bat that weak points were weekends and night hours, when senior management were away from the site; in the event of a serious accident there would be nobody immediately available to take overall charge of the site. So a control room was established adjacent to the police office and main gate, fitted with instrumentation for monitoring air contamination. A rota of personnel was organised so that there would always be an experienced senior scientist or engineer manning the control room in case of a serious incident. These people were known as Site Incident Officers (SIOs) and their duties were over and above their normal work.

I was an SIO and on my very first period of duty, at 4pm one wet Saturday afternoon, the duty sergeant showed Alexander Fleck into my office. He was accompanied by C. F. Kearton, a member of the Fleck Committee, later to become Lord Kearton. They explained that they were driving round, visiting UKAEA sites out of hours and without appointments, to find out what plans existed, if any, for dealing with a site emergency. They asked a number of questions the most probing being as to what training I had received so that I would act efficiently in an emergency. This was a fair point; indeed, SIOs had not received any specific training for the job. However I bluffed it out by mentioning that I had had the finest training possible — I said that I had once been a plant manager with ICI at Billingham works. Since they were both ex-ICI they knew what experience I would have had and hence we got on well together. On the following Monday, Cockcroft thanked me for holding the fort and said that Fleck had informed him that Harwell was well looked after in off-duty hours! But I said we had only been lucky because of my earlier experience with ICI and some thought needed to be given to SIO training

In December 1957 the first Fleck Committee report was published, on Authority Organisation (Cmnd 338). In the following month, the second report was published, on Organisation for Control of Health and Safety in UKAEA (Cmnd 342). One recommendation was for the UKAEA to increase the number of senior managers in its Industrial Group. Many of these new posts were simply filled by promotion of UKAEA staff of lower rank, then recruiting from outside to fill the resulting vacancies. But this missed the important point of the need to train the existing managers in developing communication skills and defining staff responsibilities in decision making.

One final note: after showing Fleck and Kearton out, I noticed policemen collecting revolvers from the sergeant and when I asked him what was going on, he said it was routine for the men to wear loaded revolvers on patrol after dark. I asked who they might be going to shoot at and after giving me a look of surprise he said: 'Why sir, the IRA of course!' So the police guarding atomic sites were armed after dark as far back as 1957, and prepared to take on the IRA. This was before the Fleck Report. If only the UKAEA had been as well trained and prepared to take on the dangers of radioactivity as they were to cope with a hypothetical attack by the IRA back in those dim and distant days, then the effects of the Windscale fire might have been much less — indeed, it might never have happened!

Part 2

# THE HISTORY OF NUCLEAR POWER AND REPROCESSING

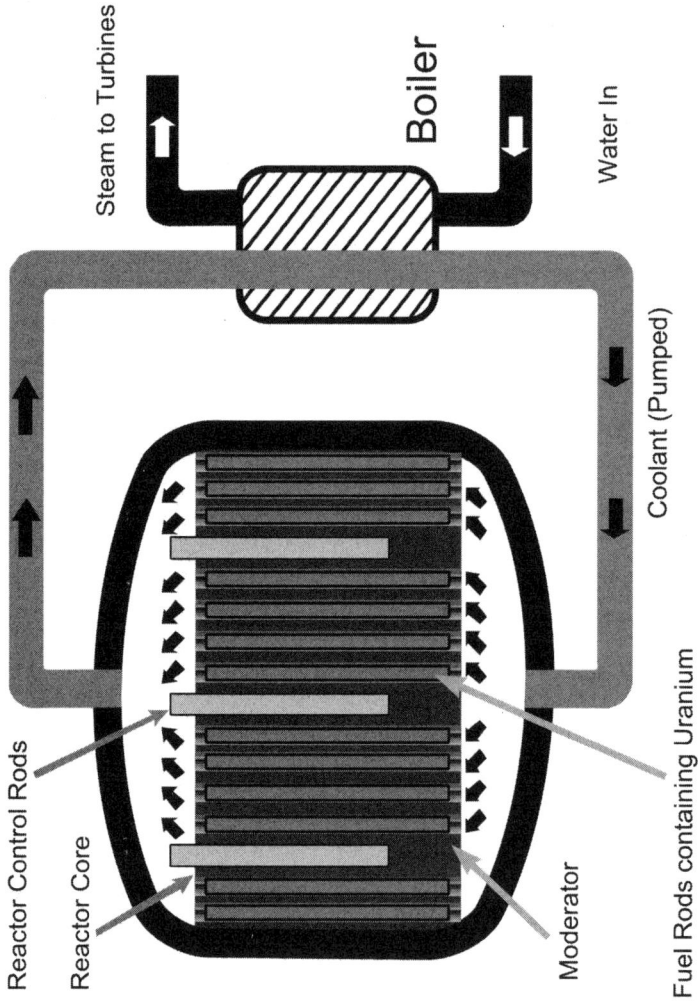

**Schematic of Nuclear Reactor**

# Chapter 11

# Calder Hall and the Magnox stations

O n a sunny October day in 1956 the Queen switched on Calder Hall, the world's first nuclear power station, and henceforth its electrical output was fed into Britain's power grid. The Queen stated that nuclear power was our answer to the energy crisis but she omitted to say that we had also built Calder Hall for the production of plutonium for use in atomic weapons. The celebratory volume *Calder Hall*, written by K. Jay and published by Methuen in 1956, stated: 'In February 1953, Churchill's government accepted a recommendation that a single PIPPA-type (Pressurized Pile for Producing Power and Plutonium) reactor should be built to produce military plutonium and electrical power'.

The design of a PIPPA reactor had been developed at Harwell prior to 1952 by two British engineers, B. L. Goodlet and R. V. Moore. They chose graphite for moderating the energy of the neutrons, gas for extracting the heat energy liberated and natural uranium for the fuel rods. In these respects the basic features were the same as those of the two plutonium piles already operating at Windscale to produce the military plutonium for the Aldermaston weapons establishment. But the main difference with PIPPA was the use of carbon dioxide ($CO_2$) instead of air as the coolant. The hot air from the Windscale piles was simply exhausted to the atmosphere via their tall chimneys, whereas the less reactive CO2 from the core of a PIPPA was circulated through a heat exchanger to generate steam, which then passed through a turbine to generate electricity as in conventional electric power stations. The cooled CO2 was then fed back to the reactor core by a circulating blower.

Building work started on the first PIPPA at Calder Hall in August 1953 and by 1956 the station had twin reactors producing 100 MW of electrical energy for the national grid whilst plutonium accumulated inside their fuel rods. Calder Hall A was closely followed by the B station on the same site at Windscale, then the identical Chapelcross A and B stations were built in Dumfriesshire, giving a total of eight dual-purpose reactors operating in

Britain. The spent fuel rods from both Calder Hall and Chapelcross were discharged and transferred in large, shielded 'flasks' to water-filled cooling-off ponds, in the case of Calder Hall on the adjoining Windscale chemical reprocessing site. When the radioactivity present in the fuel rods had sufficiently decayed, all rods were processed to yield military-grade plutonium through the same chemical separation and plutonium purification plants being used for the rods from the original air-cooled military piles.

The British chiefs of staff were very pleased with the decision to build the dual purpose stations because of the obvious chance to increase considerably the rate of output of military plutonium. They also believed that this would increase British influence on US policy in peacetime as well as providing an asset of great value in time of a major war..

## The Magnox stations

In 1954 a Treasury working party under Burke Trend reported on the economic feasibility of a civil nuclear electricity programme for Britain, independent of these Calder Hall dual-purpose stations. At the time, there was some concern for the adequacy of coal supplies for the future. Trend recommended a programme of nuclear stations which would provide an installed capacity totalling 1,700 MW by 1965 but admitted that the stations might not become economic until the later stages, if at all. Only a fairly large credit for the plutonium produced as a by-product could make the programme even approximately economic.

The Treasury working party recommendations formed the basis of the government's 1955 White Paper (Cmnd 9389), *A Programme of Nuclear Power*. This paper estimated the cost of nuclear electricity at 0.6 pence per unit but it assumed a plutonium credit of about 0.35 pence and thus the real cost of electricity would be about 1.0 penny per unit (pre-decimal money at 1955 prices). The stations were to be based generally on the design used for Calder Hall but since the uranium fuel rods would be clad in a special new alloy of magnesium known as Magnox, they came to be known as the Magnox stations. This alloy was developed for containing uranium at higher temperatures than was used with aluminium cladding and since it had a lower absorption of neutrons, it could be made thicker and hence stronger. It was the invention of Roy Huddle, a metallurgist working at the AERE Harwell.

In 1955 most of the electricity supply in England and Wales was provided by the Central Electricity Authority, soon to be transformed into the Central Electricity Generating Board. Civil nuclear power stations would of course need to be part of the supply authority. However the CEA was given little time to react to the White Paper before it was presented to parliament; the choice of reactor had already been made by the United Kingdom Atomic Energy Authority (UKAEA, or AEA). Approval for the first civil nuclear construction programme was soon given by the Conservative government, which was still led by the elderly Winston Churchill (replaced in April by Anthony Eden). In July there was a general election and the Conservatives increased their majority to 60.

The UKAEA had come into being in July 1954. Its financial and administrative powers were considerable and its control by parliament somewhat tenuous. Indeed, the UKAEA seemed to be able to exert its influence in the 'corridors of power', leading to what became known as the 'nuclear lobby'; a change of prime minister made no difference to this. Although the UKAEA was financed by direct 'vote' out of public funds, little opportunity was given to MPs to find out just what the money would be spent on. The first annual budget was considerable, being over £50 million at 1954 prices and it continued to be large for many years.

Despite the high capital costs of the Magnox stations which worried the CEA a good deal, nevertheless the UKAEA exerted influence to get the nuclear generating capacity target increased to 3,200 MW. After the Suez fiasco in late 1956 (see Chapter 7), panic in government circles about the security of Britain's oil supplies from the Middle East escalated the nuclear capacity target to 6,000 MW, for completion by 1965. This was despite the fact that a Magnox station was known to cost over three times as much to build as a conventional fossil fuel station. The CEA had argued that this level of nuclear capacity would actually lead to a surplus of electricity generating supply but the nuclear lobby held sway.

In 1957, the Central Electricity Generating Board (CEGB) was constituted, with Sir Christopher Hinton appointed as its first chairman. It will be recalled from Chapter 4 that Hinton had in 1946 been granted responsibility for the design and construction of various nuclear plants on UKAEA sites and by now his experience would well qualify him to head up the CEGB when it had newly entered into civil nuclear power generation. Actually the early Magnox stations did have a military link, for in June 1958 the Ministry of Defence announced that design modifications were to be made in the reactors so that 'military plutonium could be produced if necessary.'

But Hinton's first allegiance now was to the interests and responsibilities of the CEGB and, along with its board, he soon became convinced that the commitment to a target of 6,000 MW nuclear capacity was too premature and too expensive. The Magnox reactors would not be able to compete with new coal-fired stations with respect to steam temperature and operating efficiency, and they would be costly to build at a time when interest rates were increasing. Hinton took it upon himself to lobby the government to persuade it to lower the rate of installing nuclear stations but it was not until 1960 that he achieved any success. The target capacity of 6,000 MW by 1965 was lowered to 5,000 and completion was postponed until 1968. Although he would have preferred an even lower target, nevertheless this was a significant check on the ambitions of the nuclear lobby and had he not succeeded it is quite likely that electricity prices over the coming decades would have been higher than they proved to be.

## The consortia

The Magnox stations were to be built for the CEGB by four newly-formed industrial groups called 'consortia', each of which was led by a major heavy electrical plant manufacturer. The consortium led by Associated Electrical Industries (AEI) was called The Nuclear Energy Company, in 1956 it was awarded the contract for the Berkeley station in Gloucestershire. The Nuclear Power Plant Company, led by C. A. Parsons, was given the contract for Bradwell on the Essex coast. These two contracts were awarded by the CEGB. The Atomic Energy Group, led by GEC, was given Hunterston on Clydeside by the South of Scotland Electricity Board (SSEB). In 1957 an order was placed with the fourth consortium, led by English Electric but known as the Atomic Power Group, for the building of a station at Hinkley Point in Somerset.

The UKAEA provided design data and held courses to train consortia staff in nuclear engineering. Each consortium came up with its own variant of a Magnox reactor although the basic design had been established by the AEA. Maybe some variation was to be expected since the idea behind the setting up of so many groups was to enable competitive tendering. But the eventual size of the construction programme was not big enough to support so many competitors in the market, particularly when a fifth consortium, Atomic Power Constructions, was formed in 1959; it was given the contract for the Trawsfynydd station in North Wales.

It was in the following year that Hinton finally succeeded in persuading the government to reduce the size of the civil programme. Something had to give as far as the consortia were concerned. The first two, Nuclear Energy and Nuclear Power Plant., merged under the name The Nuclear Power Group (TNPG). Then towards the end of 1960, Atomic Energy Group and Atomic Power Constructions combined to form the United Power Company.

The first four stations all failed to meet their scheduled completion dates; Berkeley and Bradwell overran by some 20 per cent whereas Hunterston and Hinkley Point (both UPC) were about two years overdue. This set a pattern for the future. Although electricity demand was not great enough to require that these stations be completed on time, nevertheless interest had to be paid on the tied-up capital. This was a considerable sum and of course it fed through to electricity consumers. But there was a further problem with seven of the eight CEGB Magnox stations; because of internal corrosion they had to be operated at progressively lower power levels than originally designed for. Thus the output of electricity was lower and yet the overheads had still to be carried by consumers' bills.

A Magnox reactor produced comparatively little heat per unit volume of the core and so the nuclear engineers of the consortia steadily went for bigger cores in order to increase heat output. But the core was enclosed in a welded steel pressure vessel to contain the heat collecting gas. As the vessel size increased, so did the technical difficulties of construction to meet the high standard of welding necessary. Hence the seventh CEGB Magnox station, at Oldbury on the river Severn, had two reactors built with pressure vessels made of prestressed concrete in place of steel. But the final Magnox station, Wylfa on Anglesey in North Wales, pushed the design beyond current feasible limits of development. Although it was planned to be in service by 1968, it was not ready for start-up until 1971, and then it took a further four years to commission.

## The end of Magnox?

By the early 1960s, it was obvious that Magnox had no future, and no more of the genre would ever be planned. Leslie Hannah, as sponsored historian for the CEGB, was clear about the relative operating costs of electricity from Magnox and coal-fired stations during this period; the former would be a third more expensive. One reason was the decline in value of the so-called plutonium credit, which had knocked off a third of Magnox electricity costs in 1955, and was by 1958 considered likely to contribute perhaps only a fifteenth or a twelfth of production costs. By the late 1960s its contribution was going to become negative since no economic use could be found for the waste plutonium and the reprocessing of used fuel rods was costly.

From March 1962 to February 1963, the House of Commons Select Committee on the Nationalised Industries held hearings into the activities of the electricity supply industry. According to its report, published in May 1963, Sir Roger Makins, chairman of the UKAEA, had admitted that the Magnox reactors were neither competitive nor economic. Sir Christopher Hinton was even more emphatic on this score in his remarks to the Select Committee; even before taking into account the extra costs due to late completion and lower operating output levels, Magnox was already totally uncompetitive with other types of generating plant. In his evidence to the Committee, Sir Dennis Proctor, permanent secretary at the Ministry of Power, readily accepted the CEGB's figure of £20 million a year additional cost of generating by nuclear rather than conventional stations. The extra capital cost, £360 million for seven Magnox stations between 1962 and 1968, was, he claimed, too large for the generating industry to bear and added that it was 'doubtful they would have supported it if they had been a perfectly free agent'. But indirectly the Treasury would be footing the extra capital costs because of 'national policy laid down by the government', the argument being that the cost had to be borne now in order to gain a long-term advantage. Implicit was the view that nuclear power would become competitive in the future, but how could they be so sure? They were merely reflecting a widely held general view, no doubt originating from the 'nuclear lobby', that things were bound to get better in the future.

Finally, although this was the end of the line so far as the design and construction of Magnox stations went, they were all destined to run beyond their specified operating lifetime which was originally fixed at 20 years. The CEGB wanted to keep them running as long as possible because of their very high initial capital costs and would go on as long as the Nuclear Inspectorate would continue to give them a licence. But some engineers would argue that it is not possible to safely inspect all the potential danger points in very old reactors because of the high levels of radioactivity which have accumulated inside them. And this brings us to the crunch point — what happens to the Magnox reactors when they are permanently shut down? They will have to be decommissioned and this fact had hardly been taken into consideration at the time the stations were being built. We return to this important subject in the final chapter.

## Why the Magnox decision was wrong

Four main criticisms can be levelled against the launching of the Magnox programme in the 1950s and these can be briefly summarised thus:

(1) The adoption of a design for civil power generation which relied for success on selling a by-product which had a limited and unpredictable market was wrong in principle. Both industrial and domestic consumers of electricity are at any time entitled to the cheapest unit cost, on an ongoing basis, and

they were relying on the Generating Board to meet this requirement as a first priority and not have it jeopardised by outside influence from a government which was pursuing policy in an entirely different field.

(2) The programme was launched in a rush and without adequate time for considering other alternatives or indeed the basic wishes of the CEGB.

(3) A chosen prototype should have been developed to a sufficient level of definition that all potential technical problems could have been eliminated before tendering was invited.

(4) It was a mistake to build up an ambitious programme of construction with too many consortia and to an unrealistic time scale.

# Chapter 12

# The Mutual Defence Agreement

The dual-purpose reactors at Calder Hall and Chapelcross produced plutonium for nuclear weapons and were always acknowledged as military
reactors whereas the CEGB's Magnox reactors were avowedly for civil use
only — to generate electricity for the national grid. All the reactors, however,
produced spent fuel rods which contained plutonium and irrespective of
whether they came from civil or military (dual purpose) reactors they were
sent to Windscale for chemical processing to extract the plutonium.

## 'Civil' and 'military' plutonium

For security reasons, the Ministry of Defence put a blanket of secrecy on
the amounts of plutonium thus produced. The plutonium output was deemed
either 'civil' or 'military' and given a book value accordingly. The distinction
was based principally on the source but there was also a technical difference
to be taken into account. Plutonium is a man-made fissionable element which
is composed of several isotopes, mainly the desired plutonium-239 but also a
little plutonium-240 and even traces of higher isotopes. Under prolonged irradiation in the reactor, plutonium-239 begins to transmute into these other
isotopes which are undesirable for military purposes. Hence the fuel rods in
a reactor optimised for military-grade plutonium as against electricity production must be withdrawn at an early stage to avoid unacceptable build-up of the
undesirable isotopes. The lower grade 'civil' plutonium was vaguely considered to have potential for fuelling some future type of reactor not yet
developed. The UKAEA had the so-called Fast Breeder Reactor (FBR) in
mind (see Chapter 14).

## Renewal of co-operation with the US

In 1946, the McMahon Committee had not accepted that Britain had any
special relationship with the US and therefore could not expect any special
treatment. It was not until June 1955 that some opening up of the Act took

place under what was called Eisenhower's 'Atoms for Peace' programme. The US wanted to catch up on civil nuclear expertise from outside and the programme was to allow some degree of exchange of information. But the amendment of the McMahon Act was also to enable the US to co-operate in limited ways with allies in the military nuclear field. An early outcome of the amendment was a series of 'Military and Civil Atomic Co-operation Agreements' signed with Great Britain on 15 June 1955. In June of the following year the Agreements were broadened and the military and civil aspects began to become mingled. It was agreed for example that Britain might obtain equivalent amounts of enriched uranium from America in exchange for 'depleted' uranium from our reactors; a sort of Aladdin's lamp trick, new lamps for old. The enriched uranium could be used for fuelling Britain's nuclear-powered submarines (see Chapter 22).

Following the general deterioration in international relations in 1956-7 already alluded to in Chapter 7, the British and American governments evidently felt that the bonds of atomic co-operation should be strengthened even further. The British Prime Minister, Harold Macmillan, led a mission to Washington in 1957 which included the chairman of the UKAEA, Sir Edwin Plowden. They had detailed talks with the American president, Dwight D. Eisenhower, the chairman of the US Atomic Energy Commission, Admiral Strauss, and other high level officials. At the conclusion, a joint Declaration of Interdependence was signed but being subject to ratification by Congress, the necessary change to the US Atomic Energy Act did not become law in the US until June 1958. This law allowed for nuclear weapon designs and fissionable materials to be passed to countries 'that had already made substantial progress in the development of such weapons'; it was made clear that only Britain fulfilled this condition. The exchange of much classified information was clearly permitted and civil nuclear applications were of secondary importance only.

Almost coincidental with this agreement, the British Ministry of Defence issued the following statement:

'In order to provide insurance against future defence needs, certain of the civil nuclear power reactors now in the early construction or design stage are being modified so that the plutonium produced as a by-product is suitable for use, if the need arises, for military purposes. These modifications will not delay the construction of the reactors and will not affect their normal operation as civil power stations. This decision does not affect power stations at Bradwell, Berkeley and Hunterston where construction and installation are already well advanced. Hunterston is already designed as to be suitable for this purpose anyway.'

The modifications referred to in this statement were mainly concerned with fuel loading and storage provision to allow for an increase in throughput required by the military plutonium cycle of operation. The outcome of the MOD announcement was that the magnox reactors were to be optimised to favour the military plutonium production cycle and so the CEGB so-called civil stations could easily become military like Calder Hall or Chapelcross. Indeed the CEGB stations did produce 'military' plutonium as we shall see below, although in the event, only Hinkley Point A station reactors received the modifications referred to in the MOD statement. One reason for this was the switch from an A-bomb stockpile to H-bombs following the successful explosion of a British H-bomb in May 1957. Although we traded our plutonium for American tritium, a major constituent of H-bombs, the net plutonium demand was significantly less than envisaged back in the 1950s. This arrangement became possible through an amendment of the Mutual Defence Agreement, the details of which we now spell out.

## The Mutual Defence Agreement (Amendment)

The June 1958 amendment of the US Atomic Energy Act led in due course to other agreements between the British and US governments. First of all, the President signed the Anglo-American Mutual Defence Agreement in July. This was formally amended in May 1959, putting in place the legal framework for the barter arrangements in which we got the tritium and the enriched uranium from the Americans in return for British plutonium. When the British government released the White Paper containing the text of the Mutual Defence Agreement (Amendment), to give it its formal title, the political correspondent of the *Daily Telegraph* wrote, aptly and concisely:

'The immediate object is to enable both countries to produce nuclear weapons more cheaply than hitherto, by avoiding costly duplication of production capacity.'

This was on 8 May and on the same day, *The Times*' science correspondent clearly summed up the position now reached as follows:

'The most important technical fact is that plutonium of civil grade — such as will be produced in British civil nuclear power stations — can now be exported for use in nuclear weapons albeit in the USA... From an international point of view, it now seems that any country with civil nuclear power stations and also a plutonium separation plant will be able to make a nuclear weapon without need to design or operate their designs for specifically military production.'

The grade of plutonium sent to the US may have varied in composition. The early specifications for military use were very exacting; certainly anything containing more than 7 per cent Plutonium-240 would hardly have been acceptable. Also, the radioactive fission products generated whilst in the reactor had to be reduced to minute levels, which increased the chemical processing costs. But technical developments in bomb production allowed for less stringent specifications. In 1977, a bomb made from ordinary civil reactor grade plutonium was successfully detonated in the US and a scientist from the important Livermore Laboratory commented that a country could build an entirely credible national nuclear explosives capability based on only reactor-grade plutonium. All that would be required would be an off-the-shelf nuclear power station and a separation process on a large laboratory or pilot plant scale.

Only after a long silence did the subject of the fate of the exported CEGB plutonium arise in the House of Commons. It had been ignored by successive Labour and Conservative governments until in 1976 the Under Secretary of State, J. Moore, in response to a question in the House, claimed that the pluto-nium in question had gone mainly into Fast Breeder Reactors in the USA and none of it into military weapons. Then in August 1982, the Chief Press Officer of the UKAEA claimed in a letter to the press that 'none of the pluto-nium from the CEGB or SSEB Magnox reactors was exported, employed or applied for any atomic weapons or weapons-related Research and Development use in the UK, the US or any third country.'

This was just too much for Dr R. V. Hesketh of the CEGB who broke ranks and jointly with Sir Martin Ryle at Cambridge carried out some investigations into the fate of the plutonium in question. They concluded that civil pluto-nium had been exported from the UK to the US under the Mutual Defence Agreement (Amendment) during the 1960s and furthermore that it had been used in nuclear weapons. They published their conclusions and their views on the subject at some length in *The Guardian* of 19 August 1982. They finished by making two proposals for the future; a far more clearly defined distinction between the civil and military use of plutonium, and a more open system of plutonium accounting and the public scrutiny of it.

But of course the MOD would want to continue to keep all details about the fate of British plutonium under wraps. However it had been confirmed that Britain had received both enriched uranium and tritium from the USA for nuclear submarines and H-bomb production in exchange for British pluto-nium exported to the US under the Mutual Defence Agreement (Amendment). Five years after the enactment of the Mutual Defence

Agreement (Amendment), a defence White Paper (Cmnd 2270) issued in February 1964 stated that supplies of fissile material already to hand or assured were sufficient to maintain Britain's independent deterrent and to meet all foreseeable defence requirements. Production of highly-enriched uranium for military purposes at the UKAEA's Capenhurst site (see Chapter 15) had ceased in the previous year. But there was still work for Capenhurst's diffusion plant to do. In December 1965 the Ministry of Technology announced in the Commons that the Capenhurst plant was to be reactivated to supply low-enriched uranium for fuel elements for the second nuclear power programme; the AGR reactors (see Chapter 13). Then in March 1967 it was announced that 'further expansion' of the Capenhurst plant was to take place.

In conclusion, it may be said that although this chapter has revealed a somewhat complex story it has nevertheless shown how the use of nuclear fission for weapons purposes and for electricity generation are so closely inter-twined. When planning to attain the objective spelled out by the *Daily Telegraph* above, had the British and US governments given adequate consid-eration to the awesome international implications for future world stability and peace which might result from their export/barter arrangements for nuclear materials? Unhappily one is drawn to the conclusion that they prob-ably did not.

# Chapter 13

# The AGR: Successor to Magnox

In 1957 the UKAEA decided virtually on its own that the successor to the Magnox-type civil nuclear reactor should be a 'first cousin' to it, the so-called Advanced Gas-Cooled Reactor (AGR). The AEA were so confident that the AGR design would be chosen by the CEGB for its second round of nuclear stations — it was equally confident that there would be a second round — that it invested its major research and development resources in AGR technology. But this time it was decided at the outset to build a prototype reactor and by the end of the year the AEA had received the go-ahead from the government to build one at Windscale. It would have a heat output of 100 MW, producing in turn some 27 MW of electrical power, and it was intended that a full-size AGR would have a capacity about twenty times as great as the prototype, or half as large again as was planned for the final Magnox reactor at Wylfa. Although steam was to be generated at Wylfa at 385 deg C, the AGR was aiming for over 600 deg, thus achieving a major improvement in turbine efficiency. Even a fuel production line was designed and built in the early 1960s, pre-empting any decision by the government and the CEGB on the choice of second round reactor type.

It was as well that the AEA decided to go in for a prototype AGR and an associated development programme because there were some severe — and costly — technical problems to be overcome. For example, to attain a satisfactory chain reaction at the higher temperature, more fissions must take place either by reducing the absorption of neutrons by the reactor components such as fuel cans or by using an enriched fuel. A new fuel-can development programme was undertaken with industry as well as in the AEA's laboratories, investigating the possibility of using a beryllium alloy. This cost over £10 million before failure was admitted and a special stainless steel alloy was chosen which performed well when tested in the Windscale AGR. But the catch with using stainless steel was the greater absorption of neutrons compared with Magnox alloy cans, requiring the use of enriched uranium fuel. This was of course more costly than the natural uranium used in Calder Hall and the Magnox reactors. But the most intractable problems were encountered with the use of the graphite moderator at the higher temperatures in the presence of radiation and the carbon dioxide ($CO_2$) coolant gas. The difficulties were of both a chemical and physical nature and large numbers of scientists were employed in an enormous research and development programme in the AEA's laboratories at Risley in the north and Harwell in the south. In the end, a special graphite was developed in conjunction with industry and a small amount of methane gas was included in the coolant. It was as late as the spring of 1965 before the AEA was reasonably confident that the graphite problems were solved.

The Windscale AGR began working in 1962, reached its full power in January 1963 and was then soon supplying electricity to the national grid. Its designed power output of 27 MW was in due course raised to 33 MW. Although it performed well, indeed it was not finally shut down until 1981, it was not then possible to walk away from it; it had to be decommissioned and this was initially forecast to cost the government £80 million. By late 1995, work had reached the stage of lifting out the radioactive heat exchangers but there was a great deal more work to do before the site would be completely clear and restored to its original state. Undoubtedly the prototype project was a tribute to the skill and determination of all the scientists and engineers involved but it tied up much scientific and technical manpower and capital which was thus not available for other projects.

## The selection of AGR

In April 1964 the government issued a short White Paper (Cmnd 2335) announcing its second civil nuclear power programme but gave no mention of reactor type. Tenders were invited from the three industrial nuclear power consortia for the first station, although no mention was made of the intended location until the following year when Dungeness was chosen; the station would be called Dungeness B as it would go beside the Magnox or 'A' station.

Although the AEA were pressing strongly for an AGR reactor to be built and all three consortia were working from AEA data in order to produce full-scale designs, it was by no means a foregone conclusion that the CEGB would select this type. Developments surrounding water-moderated reactors in North America were familiar to reactor engineers in Britain. In the early 1960s the CEGB had shown some interest in the Canadian-designed CANDU reactor, which used heavy water as the moderator, not graphite, and natural uranium as fuel. Information from the prototype was likely to be available in advance of the Windscale AGR. But the nuclear power world was startled in December 1963 when the Jersey Central Power and Light Company in the USA accepted a tender from US General Electric for a power station at its Oyster Creek site. This was to be a Boiling Water Reactor (BWR) nuclear reactor station and it would be supplied at a fixed price, a so-called turn-key contract, and there would be no government subsidy of any kind. Within a few months, after several further orders were received by General Electric, people thought the break-through had been made to economic nuclear electricity! But by 1967, losses of $2000 million dollars were revealed; Oyster Creek was a loss leader. But in Britain in 1964 the apparent success of US water-cooled reactor designs received much attention.

The deadline for the CEGB tenders was February 1965. These were received by the new chairman Sir Stanley Brown, for Lord Hinton had just retired. The consortia had been invited to base their designs on either an AGR or a US water-cooled reactor. In the event, all three consortia had put in an AGR design, knowing the influence which the UKAEA wielded, but The Nuclear Power Group (TNPG) also submitted a BWR design — it was known to have some link with US General Electric — and the United Power Co offered a pressurised water reactor (PWR). Out of these tenders, the CEGB shortlisted an AGR and the BWR for a detailed comparative appraisal. The AGR design they chose for the study was tendered by Atomic Power Constructions (APC), which had been recently weakened when GEC pulled out of it. APC had initially decided not to submit a tender but accepted a contract from the AEA to study the design of a new AGR fuel cluster consisting of 36 rods. Then, only a matter of weeks before the deadline, APC did after all decide to tender for an AGR, with the result that this design contained much less detail than the other tenders offered. The CEGB therefore compared APC's design for an AGR with the BWR proposal from TNPG.

When the CEGB published its economic assessment, *An Appraisal of the Technical and Economic Aspects of Dungeness B Nuclear Power Station*, the economics of the two reactors were very close:

Unit generating cost for AGR design... 0.457d / kWh, and Unit capital cost... £78.40 /kW

Unit generating cost for the BWR... 0.489d / kWh, and Unit capital cost... £70.86 /kW

Much has been made of an approximate 7 per cent disadvantage being charged against the BWR, called an 'availability adjustment', because it had to be shut down for refuelling. But it was by no means certain that the AGR could accomplish a complete refuel whilst running on full load and in fact it was only in the 1980s that an AGR station managed partial refuellings at 30 per cent load. The CEGB would have done better to have assumed parity between the two designs and gone on to assess the relative merits of the two consortia from the point of view of their ability to deliver the goods and to do this within a specified time limit. They simply anticipated that Dungeness B would be on load in 1970. But the Board sent their inadequate report to the politicians and left them to make the decision.

This they did and in May 1965, Fred Lee, Labour Minister of Power, announced that an AGR would be built at Dungeness by Atomic Power Constructions. He hailed it as one of the greatest technical breakthroughs ever and serious newspapers such as the *Financial Times* echoed this view. But they were not entitled to describe the decision to build an AGR which would be 20 times bigger than a prototype AGR as a huge technical success; such a description could only rationally be applied to an accomplished fact.

The formal order for Dungeness B was placed with APC in August 1965 and specified a twin-reactor station with an electrical power output of 1,320 MW. Dungeness B was to head into great trouble but before this was generally evident, in 1967 the CEGB had ordered another AGR, for Hinkley Point, and the South of Scotland Board ordered one for Hunterston. They were to be built on existing Magnox station sites and hence they were known as B stations, as with the AGR at Dungeness. Their designed output power was similar to that of Dungeness B and they were to be built by The Nuclear Power Group. By 1969 a further two AGR contracts had been awarded by the Board, one at Hartlepool on the north-east coast, close to the industrialised area of Cleveland, and the other at Heysham on Morecambe Bay. Now each of the three consortia were to build an AGR station, each working to its own individual design.

## The fiasco of Dungeness B

The problems at Dungeness B only emerged slowly and even in 1968, the managing director of APC was proclaiming AGR's great scope for development. But after work had already got under way on site, APC were still busily turning their essentially outline design into a detailed working blueprint even though the CEGB were still expecting to have the reactors on line in 1970.

The reactor design incorporated a prestressed concrete pressure vessel lined with steel. But the tight tolerances of the working plans left little room for error in building and unfortunately the liner became distorted to the extent that the boilers would not fit inside the pressure vessel; the whole of the top half of the steel liner had to be dismantled and rebuilt. Worse was to come, as the original boiler design had to be changed and the fixings, supports and connections also redesigned. But then a vibration problem arose with the coolant gas. This proved to be intractable, and one design after another failed to overcome it. There were also problems over the corrosion of bolts which had first arisen with the Magnox reactors. Derating, or lowering the reactor power output, seemed the only way forward. In the case of Magnox stations, the corrosion problem became more severe the larger the scale and the final one, Wylfa, was derated from 1190 MW down to 840 MW. The AGR station at Hinkley Point was designed to have a capacity of 1250 MW but after commissioning it was downrated to 1040 MW.

As the constructional problems developed, APC also had continuing problems over finance, management and staffing. The CEGB apparently had some reservations from the beginning about the ability of APC to tackle the contract and admitted as much in a statement made in 1967 to the House of Commons Select Committee on Science and Technology (SCST), which had been instituted by Harold Wilson in December 1966. The Board revealed that they had specified to APC that it should strengthen its resources and recruit experienced senior personnel, a concern that had not appeared in the CEGB Appraisal of 1965. Eventually, project management at Dungeness B was taken over by a reshuffled consortium known as British Nuclear Design and Construction (BNDC) which incorporated both GEC and English Electric, but APC was to continue in existence until the station was completed.

Although Dungeness B continued to fall dreadfully behind schedule, the CEGB nevertheless maintained a stubborn optimism throughout, as the Table overleaf shows. Although fifteen years overdue, Dungeness B was still not operational in 1985. The other four AGR stations which were started by 1969 all ran behind schedule though nothing like Dungeness B. The quickest to be built was Hinkley Point B which was commissioned in 1978, six years overdue.

In 1973, Arthur Hawkins, the CEGB chairman, commented to the SCST that his Board had been strongly advised by the UKAEA to select the AGR. Dr Hannah, Electricity Council historian, maintained that the series of decisions in the mid-1960s to invest prematurely and heavily in several different and inadequate designs of AGR stations represented one of the major blunders of British industrial policy. The AGR fiasco led to a large loss of planned electrical output in the 1970s but luckily for the CEGB — and Britain — that decade witnessed a slowing down in the rate of increase in demand for electricity, so in the event an excess of capacity was actually being retained. The CEGB ordered no more nuclear stations throughout the 1970s. Blame for the fiasco must be shared out as follows:

1) The CEGB, for opting for the AGR at the time they did, for selecting a weak and inadequate consortium to build Dungeness B, and for failing to make a careful appraisal of future electricity demand.

2) The Labour government must accept some share of the blame for authorising the project so quickly in 1965.

3) Finally, the UKAEA, for going ahead with a costly investment project in AGR technology and the WAGR reactor, which was bound to exert moral pressure on both the CEGB and the government to adopt that type of reactor. The UKAEA also influenced the Board to invest in several different variations in design of full-scale AGR reactor stations as a group, all based on a scale-up factor of twenty.

## HTR and SGHWR — possible competitors to AGR

The UKAEA did not universally favour going headlong for the AGR. The decision in favour was essentially made in 1957 by an influential group centred on the reactor design offices based in the UKAEA Industrial Group HQ at Risley near Warrington. As we have already stated, considerable technical problems required solving and help was sought from many laboratories at Harwell — in the Metallurgy, Chemistry and Chemical Engineering Divisions in particular. But many AEA scientists, notably at Harwell, believed that the AGR concept of reactor did not go far enough beyond the

---

### FORECASTING THE COMPLETION OF THE AGR REACTORS AT DUNGENESS B

| In CEGB Annual Report | Estimated date | CEGB description of that date |
|---|---|---|
| 1965 / 66 | 1971 | 'completion' |
| 1967 / 68 | 1971 | 'completion' |
| 1969 / 70 | 1974 | 'completion' |
| 1971 / 72 | 1975 | 'completion of erection' |
| 1973 / 74 | 1977 | 'completion of erection' |
| 1975 / 76 | 1978 | 'completion of erection' |
| 1977 / 78 | 1980 | 'completion of erection' |
| 1979 / 80 | 1981 | 'probably complete commissioning' |
| 1981 / 82 | 1983 | 'probably complete commissioning' |
| 1983 / 84 | 1984 | 'probably complete commissioning' |

Extracted from Table 2.1 of *Atomic Crossroads: Before and After Sizewell*, J. Valentine, Merlin Press, 1985.

Magnox design. They favoured the high temperature reactor (HTR) design which offered the chance of operating with steam temperatures much higher than Magnox or AGR reactors, thus leading to a more highly efficient steam operating cycle. It would, however, have taken longer to reach the full-scale design stage than the AGR.

Although research on HTR was proceeding at Harwell in 1957, pressure on the laboratories and staff to carry out work on behalf of the AGR project soon began to denude HTR of resources. The present writer was a keen supporter of the HTR system and indeed had scientists in Harwell's Chemical Engineering Division developing materials for such a reactor in collaboration with Harwell's Chemistry and Metallurgy Divisions. Risley got deeply involved in developing the AGR and took very little interest in HTR research and development work at Harwell. Morale became low among HTR project staff, a few people accepted posts in the US, and the final blow came in 1959 when the UKAEA decided it could not afford an HTR programme alongside the big effort committed to AGR.

Fortunately, other countries in Europe were interested in the possibilities offered by HTR. At Bournemouth in November 1959, I attended a European conference on the HTR reactor. It was followed by the setting up of a joint HTR development project under the auspices of what was then known as the Organisation for Economic Co-operation and Development (OECD). A large experimental HTR reactor known as DRAGON, with a design power of 20 MW, was built at a new site in Dorset set up by the UKAEA after the passing of the Winfrith Heath Act in 1957. The reactor achieved full power in 1966 and was used for the testing of novel forms of fuel elements; it was acclaimed a great success. Hence the staff at Harwell working on HTR research and development were able to continue for several years by virtue of research contracts awarded by the DRAGON project. Unfortunately the project had to be terminated in 1976 and the reactor moth-balled, because Britain had ceased its financial contribution and the reactor was on British soil.

The spent fuel from this type of reactor would have been quite unsuitable for reprocessing at Windscale and it would not have been a source of material for nuclear weapons production. One might speculate that this could have been a reason why the UKAEA decided to reject HTR when they decided to go for AGR in 1957.

There was more support at Risley HQ for the Steam Generating Heavy Water Reactor (SGHWR, see Glossary, Appendix 1). It was a totally different reactor system from Magnox, AGR or HTR inasmuch as it employed heavy water rather than graphite as the moderator and used a liquid for extracting heat in preference to a gas. There was experience of heavy water moderated reactors in Canada. A group of UKAEA engineers interested in this class of reactor managed to gain support for building a prototype SGHWR on the Winfrith site, starting in May 1963. In September 1967 the reactor became critical and was commissioned. It reached its full design power of 100 MW in January 1968 and gained some public status by being formally opened by the Duke of Edinburgh in February. The prototype performed very well and in the early 1970s was being favoured by the Select Committee on Energy as well as the UKAEA for the next reactor choice by the CEGB. Indeed, in July 1974, the energy secretary announced that the government favoured this choice of reactor system but it fell from favour as described later in Chapter 18 and all work on SGHWR in the UKAEA ceased in 1978.

# Chapter 14

# Fast breeder:
# The never-never reactor

---

$S$ir John Cockcroft, then Director of AERE Harwell, gave a lecture on nuclear power reactors to his staff in June 1950 in which he said the long-term objective would be to produce electricity at a cost not very different from that from a coal-fired station but if this were to be done on a large scale, it would require a secure and continuing supply of uranium ore at a steady and reasonable price. If this were not possible, we would have to use a type of pile called a 'breeder', for it would breed further supplies of nuclear fuel, but in this case the fuel would be plutonium. He added:

'But these piles present difficult technical problems and may take a considerable time to develop into reliable units. Their use would also involve difficult chemical engineering operations in the separation of the secondary and primary fuels.'

In August 1955, an international conference was held in Geneva on the peaceful uses of atomic energy and there it was disclosed that the very first electric power generation by atomic energy had actually taken place using a small experimental breeder reactor at the Argonne National Laboratory in the USA; it was known as EBR (experimental breeder reactor). At Geneva, the American scientists said a 60 MW (heat) reactor known as EBR II was to follow. It would be plutonium-fuelled, would not contain a moderator and would be cooled by liquid sodium. But it was also revealed at Geneva that the British were about to embark on a similar scheme at Dounreay in the north of Scotland, well away from any centre of population — for there were safety problems to be coped with which were of a higher order than those of the gas-cooled thermal reactors.

Although the DFR (Dounreay fast reactor) would have a very small core, very roughly a quarter of a cubic metre, it was to generate 60 MW of power as heat. This heat rating is over 20 times that of a conventional thermal reactor and although this sounds impressive, it can be a disadvantage because it makes the engineering more complex and too much energy per unit volume poses safety problems; unusually high heat transfer performance is necessary and thus liquid sodium was chosen because it is an excellent heat transfer medium. Another characteristic of the breeder reactor is the absence of a moderator to slow down the neutrons and hence it is called a fast breeder reactor. A further challenge to the engineers lay in the fact that the level of reactivity, or the rate at which nuclei were fissioning, could increase extremely rapidly in the core of a fast reactor, possibly even resulting in an explosion. Thus the control system of the reactor had to be of a very high order. For safety reasons, the DFR therefore had its tiny core and accompanying heat exchangers all built inside a total containment sphere about 40 metres in diameter.

Notwithstanding the known complex technical problems of fast breeders, the UKAEA commenced building the DFR in March 1955, though in its 1956 report, it did admit that there was a risk of an accident which might lead to a rapid rise in temperature which in turn might cause melting of the fuel elements and a possible escape of fission products. It was surprising to see such a statement by the AEA, but it did have to justify the existence of a large spherical dome on the north coast of Scotland and visible for miles around. It was originally hoped that DFR would be completed by the spring of 1958, but there were various delays and the reactor core eventually went critical in November 1959. It took four years to reach its design output, finally in 1963 achieving 60 MW of heat output giving 14 MW of electricity. In general the DFR gave good service as an experimental facility; though it had a number of unplanned shut-downs it continued in service for much of the time until March 1977 when Lord Hinton pressed the button to shut it down for good.

Shortly after the government announced its second nuclear power programme in 1964, it received proposals from the UKAEA for a further stage of fast breeder development. This involved the building of a prototype fast reactor (PFR), which would have a designed heat output ten times that of the DFR; i.e. 600 MW of heat which would generate 250 MW of electricity into the National Grid. It was designed along the lines envisaged for a larger, commercial-scale power station as regards the core, fuel elements and heat exchangers. The Labour government announced its approval in the Commons in February 1966 on condition that the reactor was built at Dounreay for safety reasons. Indeed it was to be built adjacent to the existing DFR. There was by now a small fast breeder reprocessing plant at Dounreay and this was an added reason for siting the PFR near to it.

Although the government's decision did appear wise on safety grounds, it did not exactly suit some members of the UKAEA, who would have preferred the PFR to be located nearer to towns and villages as an expression of confidence in the safety of the fast breeder type of station. Since the AEA now owned a site for prototype reactors at Winfrith Heath near the coast of Dorset, where the high-temperature DRAGON and steam generating heavy water (SGHW) prototypes were located, there were those who wondered at the wisdom of generating 250 MW of electricity and feeding this into the National Grid on the very edge of the network in Scotland. It was said at the time that the decision was partly a political one, since Dounreay was in the parliamentary seat of Caithness and Sutherland and a general election was looming up at the end of March 1966, the PFR would mean more jobs. At the election, which Labour won by a fairly large majority, the Liberal MP was displaced by the Labour candidate.

The PFR was expected to be on load in 1971 but this was a vain hope for it was three years late. Technical problems were encountered with the DFR and the engineers had to take account of this. Also, suitable materials had to be found which would withstand the extremes of irradiation in the core. Yet the AEA, after talks with the CEGB and the nuclear construction firms, came out with hopes of the first commercial fast reactor (CFR) station getting started maybe as soon as 1974 depending, of course, on how the PFR was to perform. But either the AEA had not listened at this meeting, or the CEGB had since changed its mind, for at a major international conference on 'Fast Reactor Power Stations' held in Britain in March 1974, immediately after PFR had gone critical, the CEGB revealed that it was less than enthusiastic about a quick move into fast reactors. Now it did not envisage any order being placed for a CFR station before 1977 or 1978; it had to have a lot more evidence about their likely reliability.

Evidence of reliability the CEGB did not get from the PFR. Its delayed start was largely due to sodium pumps malfunctioning, requiring removal from the reactor, but the basically more serious troubles occurred after the start-up. Persistent leaks appeared in the reactor heat exchangers where hot molten sodium transferred its heat to water in order to generate steam. There were three of these units and much of the time only one of them was in operation; by 1976 the reactor had only achieved a third of its design output. There were other problems also requiring investigation and indeed, by 1979, the AEA annual spend on fast breeders and associated research and development had reached £70 million. By 1984 the prototype had completed ten years of its operating life but because of low power running and the need for a good deal of shut-down time, its effective running time was estimated to have been a mere 10 per cent of possible time. We are not talking of the performance of an experimental breeder reactor such as the DFR but a proto-type; it would have been strange if the CEGB had been keen to start building a commercial breeder station in view of the evidence.

## Facing reality

We have followed the story of the DFR, then the PFR fast breeders through from 1955 to 1984. It was not a very inspiring one, the more so when it is taken into account that the government vote for fast reactor research and development over that period came to £2,400 million (at 1982 money values). Nothing daunted, the UKAEA had persistently believed there was a clear future for this type of reactor and indeed they had been designing full-scale CFRs and pushing, for several years, for one to be built by the CEGB.

But now there came a sea change in attitude when, in evidence to the Committee of Public Accounts on 2 April 1984, the Chairman of the AEA estimated that a further 25-30 years and additional research and development expenditure of £1,300 million (1982 prices) would be required to reach the stage 'where one hopes to obtain a commercial [fast breeder] station'. To this figure had to be added £2 billion for a commercial demonstration reactor and £300 million for reprocessing facilities. The cumulative figure of £5.7 billion would finally be reached after a 60 year programme of research, development and demonstration on fast breeders. Thus, in the early 1980s, we were barely half way through.

In reporting to Parliament in 1983, on *Energy Research, Development and Demonstration in the United Kingdom*, the Select Committee on Energy declared itself

'...concerned to note that for several decades Ministers have been content to rely on advice coming almost exclusively from the UKAEA about the scale of the fast reactor programme. Accordingly, we welcome the Secretary of State's decision to review the programme in depth. We regret, however, that none of the details of this review have been published.'

At that time there was discussion about a joint proposal between France, Germany and the UK to build a commercial demonstration fast reactor in each country. But the Select Committee stressed that it could see no obvious rationale for this decision.

## The weapons connection

Looking back, we need to examine why the UKAEA pushed so hard and so long for fast breeders. The AEA has been heavily criticised for its persistent heavy expenditure on this reactor, approved of course by annual parliamentary vote, but such criticism has usually only been made from the perspective of its perceived drive for nuclear power development. But, as we have so often found in this book, once again there was a link with nuclear weapons. The Authority's military responsibilities hardly ever received any discussion or criticism in parliament, the press or the literature for the simple reason of 'national security'.

From the formation of the UKAEA in August 1954 right through to the passing of the Atomic Energy Authority (Weapons Group) Act of 1973, when the UKAEA Weapons Group was transferred to the Ministry of Defence, we need to remind ourselves that the Authority was responsible for nuclear weapons development as well as civil nuclear applications; for that reason they could give the minutes of board meetings a high security classification if the chairman so wished. Sir William Penney was the AEA board member for weapons research and development from August 1954 until June 1959, within which period the fast breeder programme got under way with the building of DFR and an associated reprocessing plant. He was a very attractive and influential, indeed dominant personality; when I was invited on one occasion to go and 'educate' him about graphite technology, he did almost all the talking. He became Deputy Chairman of the AEA in 1961, then Chairman in February 1964 when the designing of PFR and preparation of the case to build it would be receiving much attention. At the time when Penney retired in October 1967, Air Chief Marshal Sir Denis Barnett was the board member for weapons research and development and remained so until 1972.

Any nuclear technology which promised to yield plutonium would have been attractive to the weapons people in the UKAEA in the 1950s. In Chapter 11 we noted the building of Calder Hall power station in 1953, primarily for its plutonium output, and that in 1958 it was decided that some of the Magnox reactors would be optimised to favour the military production cycle. So a reactor system which promised the breeding of plutonium would have been of special interest.

The core of a fast breeder can yield 70 per cent plutonium-239 but this will be contaminated with highly radioactive fission products. Although the latter can be separated out during chemical processing, there will be some contamination with the plutonium-240 isotope, which cannot be so separated because it is chemically identical. This isotope of plutonium is less predictable in its behaviour and also carries the risk of spontaneous fission. Nevertheless some degree of contamination with plutonium-240 can be tolerated for the larger bombs, but for lighter and more compact warheads, including battle-field weapons such as howitzer shells, the highest purity plutonium-239 is necessary. This was where the fast breeder came in. Its core is surrounded by a so-called blanket of natural uranium, consisting mainly of the uranium-238 isotope which becomes bred into plutonium consisting of 95 to 98 per cent plutonium-239. This is highly desirable for nuclear weapons purposes and is easily separable by chemical means from the remaining uranium.

## The doubling time factor

The appeal of the fast breeder for civil nuclear power use is two-fold. First, the ability to fuel with plutonium if uranium ore is likely to be scarce and costly in the future. Second, the chance of a net gain in plutonium during operation. This gain is commonly assessed in terms of the doubling time.

A doubling time is the number of years it takes one breeder reactor to breed enough plutonium to provide the core of another breeder, whilst still keeping itself fuelled. A figure of 30 years has been estimated as a possibility but that is not taking into account the fact that the reactor has to shut down at inter-vals to allow for reprocessing — the core to have fission products removed and plutonium fuel replenished and the blanket for the purpose of extracting the bred plutonium from the remaining uranium. Then the new fuel elements have to be fabricated as have the new blanket sections; a one-year shut down time, including the initial cooling of the extracted core, would be extremely good going. Then of course these operations unavoidably result in a loss of plutonium from the system at each shut-down, maybe as much as 7 per cent, and topping up with fuel from outside is required. It is likely that in practice, a realistic doubling time of about 60 years or even more may be anticipated. But this would probably mean a failure to provide the core of another breeder within the lifetime of the initial plant. In any case we would have to wait very many years to find out if the initial promise of plutonium breeding could be fulfilled in practice.

## The safety factor

There are a number of features regarding the fast breeder design which increase its risk of an accident compared with that of thermal reactors. We have already touched on some of these, relating to the core reactivity and related control aspects, the inherent dangers of using molten sodium as the coolant and the risk of leakage in the coolant circuits. No organisation in the world has built a full size commercial fast breeder station, in which each reactor would contain about 6,000 tonnes of sodium at a temperature of between 400 and 600 deg C. After experiencing sodium fires at the Shevtchenco fast reactor site in Russia, their technologists became very chary about expanding the breeder programme. When the British government's Secretary of State for Energy privately asked members of the Nuclear Inspectorate in January 1977 what would happen if a fast breeder were to blow up, he got the reply that about 10,000 people would be killed. However, it must be recorded that the British programme on fast breeders involving DFR, PFR and the small reprocessing plant had a fairly good safety record, as indeed the French had with their breeder development involving the building and operation of the two reactors Phenix and Super Phenix.

But in the present writer's view, based on experience in designing reprocessing plants, accidents would be more likely to take place not in the reactor itself but rather in the fast breeder's reprocessing plant because of the risk of what are called 'criticality' accidents. These are where too much plutonium unexpectedly gathers into a particular configuration and an explosion can occur. Any such happening is likely to have devastating results because reprocessing plants are not built like a reactor in an all-enclosed pressure-resistant dome.

## The end of the British fast reactor programme

The prototype fast reactor was operated until March 1994 when, after a life of 20 years, it was permanently shut down. This was because in 1988, Cecil Parkinson, the Energy Secretary, had said that funding of fast breeder development in Britain would be cut off in March 1994 because there was no prospect that these reactors would either be economic to build and operate, or indeed necessary, for at least 30 to 40 years. So, after four decades of endeavour and four billion pounds of expenditure, the project to harness the promise of plutonium breeding had failed dismally in Britain. And not only in Britain, for fast breeder reactor programmes also died in Russia, the US and Japan.

## The aftermath at Dounreay

Although the programme of a nuclear research site may be terminated, health and safety problems can and do remain for a long time. The Dounreay site offers clear evidence of this.

During 1996-1997 the Nuclear Installations Inspectorate (NII) carried out an examination of the fuel cycle area at Dounreay. The Inspectorate's report was sent to the UKAEA in the summer of 1997 but was only made public in June 1998. It was a very critical report; it stated that there was a severe lack of properly qualified personnel in critical areas of the facility, it recorded many instances of poor plant design and engineering and even said that the UKAEA had lost sight of its duties as a licensee at Dounreay.

During the early months of 1998, the Dounreay site suffered a number of safety incidents. Then in April the Americans transported four kilogrammes of highly enriched uranium and 800 grammes of irradiated fuel from Tbilisi in Georgia to the UK for safe keeping at the Dounreay site. This was the outcome of secret negotiations between the Georgian, British and US governments; it was to prevent the material from falling into the wrong hands and for safety reasons. There were conflicting statements in the media as to why Tony Blair agreed to accept the material whilst the French and Americans had declined to. Then the problem for the British government was seen to be aggravated by the resignation of the chief of the UKAEA Constabulary, who complained about conditions at British nuclear installations including Dounreay, and said he could no longer guarantee the security of installations against attacks by terrorists and environmental activists.

Having noted the growing strength of public opinion over the Dounreay nuclear site, Donald Dewar, the Scottish Secretary, finally pledged that no more contracts would be signed for dealing with any form of nuclear waste at the site and after completion of existing contracts in the early years of the next century, the reprocessing plant would be shut down. But shortly after this, it was said that Shell were hoping to remove about fifty cubic metres of contaminated pipework from the Brent Spar oil rig, moored in a Norwegian fjord, to Dounreay so that radioactive scale could be removed and stored there. At the time of writing, it was thought that the Scottish Environment Protection Agency was considering an application for an import licence.

# Chapter 15

# Diversification and reorganisation of the UKAEA

When Prime Minister Attlee and his secret cabinet committee GEN 163 met in January 1947 to endorse the plan to make Britain a nuclear power, they inevitably committed the country to establishing a network of geographically scattered though closely linked nuclear sites. This network is sometimes referred to as the 'nuclear cycle' but it is unlikely that GEN 163 would have given a thought to the likely creation of such a costly and potentially dangerous nation-wide system, all to result from their perceived need for Britain to manufacture nuclear weapons as urgently as possible.

## The nuclear cycle

Britain's nuclear cycle developed from September 1947 onwards and grew in complexity. We have already mentioned many of the establishments on the nuclear cycle network and now a visual aid may well help to clarify the picture as a whole. The simple flow diagram shown overleaf illustrates the main operations and their linkages as they would have appeared in the mid-1960s, following which a reorganisation of ownership and responsibility took place. This is described below. A further change made by government about this time was to make it legal for the UKAEA research laboratories to become involved in non-nuclear projects with industry. This is also described later. We now comment on the main features to be noted on the flow chart.

The two top blocks of the diagram are grouped together by back-shading because of the closeness of the operations, uranium ore mining and refining, which do not take place in Britain. They are discussed in Chapter 16.

Uranium enrichment takes place at the Capenhurst Works in Cheshire. In 1965, various levels of enrichment of the uranium-235 isotope in natural uranium could be carried out. A low degree of enrichment was first achieved there by gaseous diffusion in April 1953. Although modest levels are required for reactors such as AGRs and PWRs, very much higher levels of uranium-235 are necessary for atomic and thermonuclear weapons which use uranium,

## Britain's Nuclear Cycle (1965)

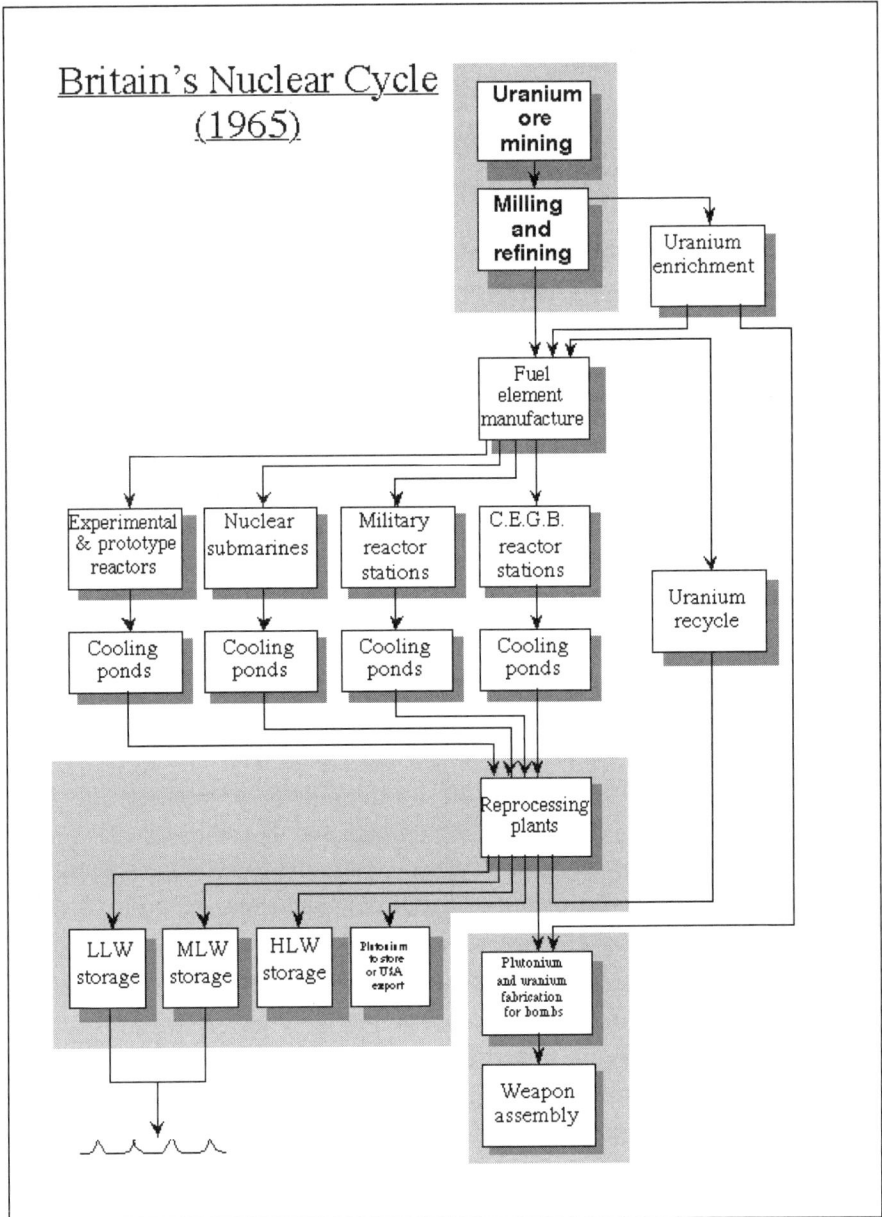

and vast amounts of electrical power are consumed in this case. The Capenhurst site must be regarded as a dual purpose civil/military factory.

Fuel element manufacture is carried out at the Springfield Works near Preston, Lancashire. The first casting in Britain of a metal nuclear pile cartridge took place at this site in January 1948 using Canadian uranium but in October casting was based on Belgian Congo ores. Later, not only natural uranium would be used but also recycled uranium from the reprocessing plants at Windscale.

The natural uranium and enriched fuel elements are transported to the various reactor stations spread around the country but on the diagram, these are grouped into four blocks for the sake of simplicity: CEGB and South of Scotland civil stations; military reactor stations at Chapelcross in Scotland and Calder Hall at Windscale, Cumbria; nuclear submarine reactors of which only a land-based one at Dounreay was operating by the mid 1960s (see Chapter 22); and experimental/prototype reactors which included the Windscale AGR, the DFR at Dounreay and the SGHWR at the AEA's Winfrith Heath site in Dorset.

All the reactors have their spent fuel elements 'cooled' in ponds, under water, to allow the original very high levels of radioactivity to decay sufficiently so that transportation to the reprocessing site can be safely carried out. The reprocessing site is at Windscale Works, called Sellafield from 1981 onwards. It was always assumed that reprocessing was essential in order to be able to extract plutonium and depleted uranium. Our diagram shows the reprocessing plants in one box and the three waste outputs and plutonium in four others, inside a back-shaded area to show their grouping is together at Windscale. The significance of the arrow in the bottom left-hand corner of the diagram is the passage to the Irish Sea of certain liquid radioactive wastes through an overhead pipeline. The development of the whole reprocessing complex is discussed later, in Chapter 17. At the foot of the diagram two boxes represent weapon materials production at Aldermaston in Berkshire, and weapon assembly at nearby Burghfield. Back-shading on the diagram emphasises the close character of these highly secret establishments, which are referred to later in Chapter 22.

At each of the sites, the occurrence of radioactive substances poses a risk to the health and safety of operating personnel and staff, and potentially to the general public living nearby. Hazardous materials unavoidably have to be transported between sites, by road and rail, thus creating a further potential risk in the event of an accident during movement. Although much thought would have been given to the *selection* of each individual site, it is evident that no thought had been given at the outset to the overall planning of the nuclear cycle *network*.

## Diversification

When Labour came to power in 1964, the new government decided that industry ought to be able to benefit from the research and development expertise available in the public sector and in particular, the huge technological and scientific resources built up by the UKAEA in the laboratories of their establishments and the nuclear cycle. So when the Science and Technology Act of 1965 was passed, Section 4 was drafted to allow the AEA to spread its activities outside the nuclear industry. It was all part of Prime Minister Harold Wilson's so-called 'white heat of the technological revolution', which slogan had formed a key part of his election campaign.

'Diversification' became the order of the day for the AEA. It came as a godsend to AERE Harwell in particular, for at that time there had been rumours of the establishment being closed down in order to release scientists and research engineers for employment within private industry. There had been much discussion about this in the media so the morale of Harwell staff was very low; their view was that 'the whole was greater than the sum of the parts'. In other words, keep the existing teams together and let them enter into joint technological projects with manufacturers. Fortunately for Harwell, the Labour government also saw it that way! A trading fund was set up in April 1965 to cover the AEA costs of joint commercial projects with industry. Research managers had to search out the opportunities and then put forward projects for approval, having first secured agreement of all the firms to share in the financing. Sir Walter Marshall, then Deputy Director of Harwell, took personal responsibility for getting the non-nuclear programmes under way. He set up a marketing office. He would tell scientists, 'You are not salesmen — you are going to market your skills and experience; learn the difference.'

I well remember the problem we had with getting the initial set of proposals approved by government. They were grouped under three heads — The Ceramics Centre, The Non-destructive Testing Centre and a Materials Advisory Bureau. The predominance of materials technology will be noted here; at that time, Harwell was probably the most prestigious materials research laboratory in Europe and it was right that its resources should be applied more widely than merely in the nuclear field. After Marshall sent the package of proposals to the newly-formed Ministry of Technology (known as MinTech) the weeks rolled by without any feedback from London. Harwell staff began to get anxious. Then one Saturday afternoon found me at a meeting in Reading's Cooperative Hall to hear Dr Jeremy Bray, a Minister from the MinTech, speak on government policy regarding technology in industry. Suddenly I heard him say:

'Those Harwell scientists up in their ivory tower, not far from here, ought to come out, stump the country and see what they can do to revitalise British industry.'

I found myself standing up to interrupt the Minister and I shouted loudly: 'We've already done just that! You folks ought to get your fingers out and approve our proposals without any further delay on the government's part.'

But Bray was shaking his head and as the chairman called for me to shut up and sit down, stewards were moving grimly towards me. But I would not give way and insisted on a point of order that the Minister withdrew his statement. I pointed out that I was a senior scientist from Harwell and I was not going to have my establishment misrepresented in this way. I just managed to say that my director had sent our proposals to the ministry before uproar prevailed. Bray sought me out at the tea interval and asked me just what Harwell had done. When I had finished answering him he assured me that neither he nor Tony Benn, the Secretary of State, knew anything about it but he promised to find out what had happened, as a priority. Four days later, an assistant director told me that Harwell had that very morning received formal permission in writing from MinTech for us to start on our three non-nuclear projects. Harwell's proposals had sat on a deputy secretary's desk all that time — apparently he had not seen anything like that before and did not known how to deal with it.

Harwell was now out of its ivory tower. Soon, other projects were proposed and approved from other establishments too. MinTech realised there ought to be a mechanism for assessing projects, to estimate the likely return on investment, and for this purpose the Programmes Analysis Unit (PAU) was set up. It was based on the Harwell site, just outside the security fence. It recruited people from the UKAEA and also civil servants from MinTech. It was supposed to be quite impartial in its judgements.

Thirty years on, the non-nuclear projects added up to a large chunk of the UKAEA's research and development activities. They had become known as AEA Technology, which brought in a modest annual profit for the treasury's coffers. In 1996 the Conservative government under John Major therefore decided to privatise it. It proved to be a difficult and very expensive task to separate it from the rest of the UKAEA, which of course did not bring in a profit and was therefore not to be put on the market.

## Reorganisation

Returning to the nuclear cycle diagram, it is necessary to point out that it is incomplete. It does not show the links with British-built nuclear power plants in Japan and Italy, inasmuch as their spent Magnox fuels were due to come back to Britain for reprocessing by the UKAEA, after which the extracted plutonium and uranium would be sent back overseas. Reprocessing contracts of this nature looked like being a lucrative business for the AEA. Clearly, the organisation saw itself entering a commercial stage of development in the late 1960s and it hoped to build up this export activity. The Labour government under Harold Wilson supported this aim and decided that some legislation was required in order to hive off the AEA's fuel-service activities as a business in its own right. A Bill for this purpose was duly drafted and got as far as its first reading in the House of Lords in March 1970. But then the government was swept from office in the general election of June 1970 and so the Bill fell. However as a consequence of Britain's bipartisan policy on nuclear matters, Edward Heath's government soon resurrected the legislation and passed it through parliament without any difficulty. The AEA, which had more than doubled its size during the sixteen years of its existence, now became broken up by the Atomic Energy Authority Act of 1971.

This Act transferred the UKAEA's Production Group to a newly set-up government-owned company known as British Nuclear Fuels Ltd (BNFL). Also, the UKAEA's Radiochemical Centre at Amersham was made into a separate company. About the same time, the National Radiological Protection Board (NRPB) was created as a result of the Radiological Protection Act of 1970. BNFL had its headquarters at Risley and although nominally independent of the UKAEA, its shares were owned by the latter. Sir John Hill, chairman of the UKAEA, was also made chairman of its two new off-shoots. BNFL was a very large chunk of the old AEA for it included the uranium and fuel-manufacturing plant at Springfields, the uranium enrichment plant at Capenhurst and all the plants at Windscale except the Windscale AGR reactor. As BNFL took with it the responsibility for producing fissile material for nuclear weapons, it was responsible for the Chapelcross dual purpose station as well as Calder Hall. So although BNFL was appointed as a commercial company it had an ongoing military client, the MOD, with which its financial arrangements would be undisclosed.

The new structure did nothing toward unravelling the civil and military tie-up in nuclear affairs, indeed the B205 plant at Windscale (which is described in Chapter 17) was still processing military and civil materials, although in August 1971 the government did announce that it intended to move the Weapons Group of the UKAEA into the Ministry of Defence. This transfer was not complete, however, until April 1973.

## Geography and the nuclear cycle

Our diagram of the nuclear cycle does not show the geographical location of nuclear sites in Britain and as we have already pointed out, their distribution is not the result of any overall master plan. The nuclear power stations are widely scattered throughout England, north Wales and south Scotland. Spent fuel is therefore transmitted long distances by rail or road to Sellafield reprocessing site on the west coast of Cumbria, which is as far from the power stations as it is possible to achieve. Sellafield, alias Windscale, was the site of an ordnance factory during World War II and was so located as to put it as far as possible from German bombing planes. Thus the shielded flasks, containing the radioactive spent fuel, have to travel many miles to get to their destination, with all the attendant risks of an accident. Furthermore, the Sellafield site is a long distance from Aldermaston in Berkshire, the destination of the processed plutonium, uranium and certain other radioactive materials required for nuclear weapon manufacture. It is always claimed by the UKAEA that the flasks have been thoroughly tested to ensure that in the case of accident, dangerous contamination cannot occur. But how thorough is thorough? It is not universally accepted that there is no risk of radioactive contamination likely to result from the collision of material being transported around the nuclear cycle in Britain.

Comparison of electricity load demand centres with nuclear reactor station sites reveals that there is much disparity here. The result is that nuclear-generated electricity has to travel further than fossil-generated electricity to reach demand; thus will the loss of energy during transmission be proportionately greater.

But the greatest danger to the British people from all the various nuclear sites would lie in the event of war and not necessarily nuclear war. Enemy bombing of the homeland would almost certainly include nuclear stations and factories. However effective was home defence, some conventional weapons would be bound to reach their targets and the extent of dispersion of radioactivity would depend on the wind direction and force. Many nuclear sites have been located on the coast, perhaps to provide a supply of cooling water or a sink for the discharge of low level liquid waste. Such sites would be easily-reached targets for inflying aircraft.

# Chapter 16

# It all started with uranium ore

The production of all nuclear weapons, both fission and thermonuclear, and the generation of nuclear electricity, have all depended upon access to a source of naturally-occurring uranium ore. (In principle, thorium might have been used, but it only occurs in concentrations very much smaller than the richer uranium ores.) Uranium metal does not occur in a natural state and ores containing it in an oxidised state, usually based on the formula $U_3O_8$, have to be mined together with the mother rock in which the uranium oxide is always present in low concentration rarely exceeding 0.3 per cent. Even at this level, a price of nearly $22/ Kg would have been demanded in 1973 in order to show a profit because of the costs of mining, extraction and concentration. For nuclear applications, whether power generation or bombs, it is only the U-235 isotope which is capable of fission and since this is only present in uranium metal at 0.7 per cent, less than 28 grammes of active uranium fuel is present in each tonne of the original ore, or 'yellow cake' as it is sometimes called in the trade because of its pronounced colour.

As there has not been any uranium mining activity in Britain there is a tendency to ignore the dangers to health of this activity; one has often heard it said that nuclear electricity production is a clean, healthy business which eliminates the sending of men down unpleasant and potentially dangerous coal mines. However when uranium ore is mined it emits a radioactive gas called radon which rapidly decays to a radioactive solid (radium D). These carcinogenic elements are often inhaled into the lungs of miners. After being mined, the ore is milled and refined. Thousands of tonnes of waste ore (tailings) are left on the ground and after drying out are liable to be blown about and can thus be inhaled by operatives and any of their families resident in the area. This dust is radioactive and uranium miners and members of their families are known to have died of cancer. Any large increase in nuclear power production could undoubtedly lead to a considerable increase in cancer deaths at the sites where uranium ore is mined unless stringent and hence costly measures are introduced and maintained.

## Obtaining uranium supplies for Magnox reactor fuel and nuclear weapons

The British Magnox station building programme got underway in the 1960s and the manufacture and development of nuclear weapons continued at Aldermaston. It was clear that increasing supplies of uranium ore would be required. This need was discussed in the Labour cabinet, discussions took place between MinTech officials and UKAEA staff and members of a large British-based international mining company known as Rio Tinto Zinc (RTZ). Although the Americans, the French and the Soviets all had their own indigenous supplies of uranium, the British did not. However there were deposits in Commonwealth countries and mines were being worked in Canada, where RTZ were active at the Elliot Lake site in Ontario through their subsidiary, Rio Algom. The initial target was 10,000 tonnes of uranium oxide, to be delivered to Britain between 1966 and 1982.

In the event of increased supplies being required, the government assumed that these would also come from Canada. But in the small print of the brief submitted to the cabinet by MinTech there was a caution that in the remote possibility of the Canadian subsidiary failing to supply, the contract would be switched to a supplier in South Africa. However the brief does not appear to mention this being in Namibia (South-West Africa), where RTZ were intending to develop the Rossing Mine for extracting uranium ore. George Brown, then Foreign Secretary, noted the small print of the brief and tried to insist that the contract should only go ahead if there was no chance of South Africa becoming involved — he was unsuccessful and the contract was signed by MinTech's secretary of state, Tony Benn, in March 1968. If the small print on it had been queried and the UKAEA had been pressed to seek amplification and clarification of it with RTZ, it is unlikely that Tony Benn would have agreed to sign the contract.

## Britain defies international law

In 1948, the Afrikaner dominated Nationalist Party in South Africa inaugurated a legislative programme for the social, economic, and political separation between white and non-white members of its population. This was called apartheid. South Africa held a mandate to administer South-West Africa, later Namibia, but she was required by the United Nations to bring that country toward independence. When it became clear that she had no intention of doing so and furthermore rejected out of hand the right of the UN to interfere, the International Court of Justice, at the request of the UN General Assembly, ruled in 1950 that the UN was the proper supervisory power. South Africa ignored this ruling and began imposing apartheid on the Namibian people. Eventually, in 1966, the UN revoked South Africa's mandate over Namibia and established the UN Council for Namibia as the sole legal administering authority for the territory. In August 1969, the Security Council gave full backing to this arrangement and called on South Africa to withdraw from Namibia at once, but she refused. In June 1971, the International Court of Justice declared South Africa's presence in Namibia as 'illegal' and instructed that country to cease occupying Namibia; the ruling was ignored. The General Assembly of the UN then adopted a decree for the protection of the natural resources of Namibia; this gave the Council for Namibia the right to seize any minerals wrongfully exported.

Incredibly, although in 1969 the Security Council instruction to South Africa to withdraw from Namibia had been refused, in the following year the planned supply of uranium to Britain was switched by RTZ from the Rio Algom mine in Canada to the Rossing Mine in Namibia. The UKAEA knew about this and apparently did not dissent, maintaining that MinTech had been informed. However, the secretary of state said the information did not reach him and thus there was a conflict of views over the exact circumstances regarding the switch from Canadian to Namibian ore supplies. What is clear is that in view of the facts given in the previous paragraph, by going on with the intention of purchasing uranium from RTZ knowing full well that it was coming from Namibia, the British government was intending to breach international law and flout the will of the UN Security Council and General Assembly.

There was quite a row between the civil servants involved and members of the cabinet. Shortly afterwards, the Attorney General, Sir Elwyn Jones, ruled that a *force majeure* clause in the contract safeguarded any policy decision which the government cared to make and so Lord Brockway called for the contract to be cancelled forthwith. But Lord Lovell-Davis, speaking for the government, claimed that there was a world-wide shortage of uranium and Britain would not be able to obtain uranium from any other sources during the period of the Rossing contract. He was leaning on briefs from civil servants but it is clear that they were either keeping information back, or they had not done their homework, for there was plenty of uranium available at the time from within the Commonwealth. (South Africa had withdrawn from the Commonwealth in 1961 because of its 'disapproval' of apartheid.) Wilson declined to take action and instead resorted to the ploy of setting up a 'secret inquiry', the outcome of which would be kept under wraps until after the general election. Here we have yet one more example of atomic energy matters being kept under a cloak of secrecy.

## Tory and Labour governments stick with Namibian uranium contract

When the Conservatives took office after the election, nothing was done about the Rossing contract and the RTZ subsidiary company began preparing the site for the mine in Namibia. Initially it was to be mined by the open cast or surface removal method because the ore-containing bodies were near the surface. However, according to Dr Bowie, consultant to British Nuclear Fuels at Risley in 1979, the average ore grade of the Rossing uranium ore was low, being only .035%, compared with ore in the Elliot Lake province in Canada which was about three times greater at .10%. Presumably the low labour costs in Namibia compared with those prevailing in Canada would more than offset the difference in cost of the ore grades.

During their period in opposition the Labour Party gave a good deal of consideration to the issue of the Rossing contract and in 1973 a pledge to terminate it was fully endorsed by the party conference, but apparently this did not get included in the party's 1974 election manifesto.

Labour was returned to office after the general election in February 1974 and Harold Wilson found himself Prime Minister of a minority government — which restricted him a great deal. The foreign secretary, James Callaghan, stated in the House of Commons in December that South Africa's occupation of Namibia was unlawful and that it should withdraw forthwith. He also said that the government would give no further promotional support for any trade with Namibia. Finally he promised support for the international community to bring about the withdrawal of South Africa as called for by the United Nations.

Notwithstanding this declaration by its foreign secretary, the Labour government still did not cancel the contract for uranium from the Rossing Mine, now developing rapidly in Namibia. Yet Wilson must surely have realised that all taxes and revenues paid by RTZ's Rossing Uranium subsidiary would go to the South African government. And so this blatant support of the South African regime, illegally occupying Namibia and pursuing its policy of apartheid, continued through the life of the Labour government and that of the Conservative administration which followed it later in the decade.

James Callaghan took over the premiership when Harold Wilson decided to retire at the age of 60 in 1976. When asked in a press interview whether the troubles in South Africa could threaten supplies of key raw materials from that area he is reported to have replied: 'Any prudent country ought to be looking for alternative supplies.' Whether as a result of the PM's observation or not, the CEGB did set about finding other sources of uranium ore off its own bat. After all, they would be the main benefactors, ultimately, of the Rossing ore and any interruption in its supply could affect their nuclear electricity programme. The Board did find other supplies were available, from Australia and Canada and also Niger, thus demolishing the government's argument that Britain had no choice but to depend on RTZ's Rossing mine. In January 1979, the Prime Minister of South Australia visited Britain and took the opportunity to inform the energy secretary, Benn, that his State had about 250,000 tonnes of uranium with a market value of £1.25 billion.

## The unanswered questions

There are a number of questions which still need answering regarding Britain's uranium requirements after 1968, the scheduled completion date of the Magnox civil reactor building programme:

• Was there at any time an imminent shortage of uranium ore for producing fuel rods for the military and civil reactor operations?

• What was the size of the British stockpile of uranium over the period?

• Exactly where did all our supplies of uranium ore come from?

These matters of fact would not have been tabled under the excuse of national security ('the Soviets could work out how many weapons we might be making').

But there are other questions which also need answering concerning uranium supply and which hardly raise security issues:

• Where did responsibility lie for deciding what was needed? The UKAEA? The CEGB? The government?

• Was there a clearly spelt-out set of rules governing the invitation and placing of supply contracts for uranium?

• What were Britain's relations with the Australian and Canadian governments over possible supplies of uranium throughout the period being studied here?

• Finally, why did the Labour government approve a contract with RTZ which it believed would bring supplies of uranium ore from an operating Canadian mine when it became apparent that the contract meant ore from the

Rossing Mine and the latter not yet started? Furthermore, why did the Labour government still refuse to withdraw from the Rossing contract even when not to do so was clearly breaking international law and was flouting the will of the UN Security Council? And why did the succeeding Conservative government behave likewise?

A little light is thrown on some of these matters by studying an extract from the published diaries of the man who was the Labour government's Secretary of State for Energy, Tony Benn, during this period. It refers to a meeting on 28 April 1977 of a committee known as GEN 74, which dealt with nuclear policy, nuclear exports etc. and was where 'you get a ministerial view forming against that of the officials and, in particular, of the nuclear lobby' (Benn). The first paper was on uranium demand and supply, put in by Benn's own department. He had referred back the first draft because it had excluded what he had asked for, namely a comparison between forecast demand for uranium and forecast supply, particularly over the previous four years. The second draft was full of incomprehensible figures but it was nevertheless presented to GEN 74 that afternoon and Benn succeeded in drawing attention to key figures to draw out his conclusion: that although forecast demand had far exceeded forecast supply in the first half of the 1970s, with the cut back in nuclear installed capacity the forecast supply now exceeded forecast demand, which meant that availability of uranium was not as tight as anticipated. 'This', he said, 'absolutely destroyed the officials' case. We did have time,' he said, 'that was the important thing.'

# Chapter 17

# Reprocessing problems, THORP, and the Windscale strike

Windscale Works was the site of a Ministry of Supply ordnance factory known as Sellafield during the 1939-45 war. Then in 1947 the Ministry's Department of Atomic Energy (D At En) took it over as a nuclear establishment. Construction of the ill-fated Piles 1 and 2 began in 1947 (see Chapter 4). A pipeline to discharge low-to-intermediate radioactive liquid wastes (LLW and ILW) into the Irish Sea was laid in June 1950 and in the following June a plant for the storage of high level radioactive waste (HLW) was commissioned. In January 1952, the new plutonium separation plant (Building 204, usually referred to as B204) began to process the first batch of irradiated slugs (an early term for fuel rods) which until then were stored under water after removal from Pile 1. This initially seemed to go well, but problems arose as described in chapter 6 under Operation Hurricane. Just over the river Calder in August 1953, construction of Calder Hall power station began and was completed in 1956 (Chapter 11). In due course, the irradiated fuel elements from this dual civil/military station came back over the river for plutonium separation and later, the elements from the Chapelcross dual-purpose station were also fed into Windscale.

When the CEGB programme for building Magnox reactors was started (Chapter 11), it became clear to the UKAEA that it would produce more spent fuel than the original separation plant B204 could possibly handle. Magnox fuel stayed in the reactor longer and thus received a higher burn-up of fissionable material than occurred with reactors built primarily for military use as at Calder and Chapelcross. This made the additional spent fuel more radioactive too. So without much ado, the AEA designed and built another reprocessing plant adjacent to B204. In June 1964, after a year's delay, B205 came into operation. Another plant, B30, was simultaneously commissioned for the purpose of stripping the Magnox cladding off the fuel rods.

Very little information was published about B204 or B205. Because they both produced plutonium essentially for military purposes, some of it involved in deals with the US military complex (Chapter 12), a high security classification was felt to be justified. The plants also produced plutonium ostensibly to store for civil fuel use, nominally for fast reactors at a later date (Chapter 14). Just how much was stored has never been published. But there was a further complication inasmuch as B205 also received spent fuel returned from the British-built Magnox-type reactors at Latina in Italy and Tokai Mura in Japan. *Even though separated plutonium was potential nuclear weapon material, it was still sent back to these two avowedly non-nuclear weapon owning states.*

## The downside of reprocessing

An inevitable result of developing a large reprocessing site like Windscale is that after a couple of decades one finishes up with a complex of radioactive process vessels, storage tanks and a piping network, all radioactive to some degree. So any leak, spill or pumping mishap can lead to the contamination of a building, the substrate or even, via the pipeline, the Irish Sea. Contamination occurred on many occasions at Windscale and sometimes workers were even contaminated in the process of digging to find the source of leaks, as for example in 1976 in the case of the leak of HLW from a waste-storage silo in Building 38. This leak had been discovered quite by accident during excavations near the silo. Radioactivity levels continued to increase in nearby soil because it seemed that no one knew the exact source of the leak and how to stop it. This leak gained some notoriety because news of it failed to reach the Energy Secretary, Tony Benn, for two months or so, much to his annoyance given his avowed policy to create a more open attitude about incidents at Windscale.

BNFL continued to drill exploratory holes in order to be able to assess the subterranean radioactivity profile at the Windscale site. In March 1979, in one of their boreholes they found HLW which they considered could not possibly have come from a leak under B38. The source turned out to be a small disused building which had been used at one time to tap off samples of HLW for transport to Harwell for research into the encapsulation of liquid wastes in solids. *But the latter work had been terminated in the mid-1960s because the UKAEA Production Group had said there was no need for any such research programme. The writer can well remember the anger of the Harwell scientists at that time; they felt they were doing very important work.* Sometime during the next ten years, a leak of HLW developed in the disused building and as no one ever went to check out the building, the ground below and around it became highly radioactive, perhaps to the tune of 30,000 curies. Within a month or so, this leak had been found and dealt with but three months later, BNFL were again in trouble. There was a fire in B30 in the section where the coating of Magnox fuel was removed. Although it was not very serious and the plant was only shut down for a few days, several workers were contaminated. The general background radiation in B30 was thus increased, bringing forward the time when a replacement plant would be required.

But by now, the Nuclear Installations Inspectorate (NII) said it was time for them to make a study of safety conditions at Windscale. After the general election on 3 May 1979, which the Conservatives won, Margaret Thatcher became Prime Minister. Approval was given by the new government for the NII to go ahead with the inspection. But it was not until February 1981 that the government received the report. In it, BNFL were severely criticised over conditions at their Windscale site — for plant design weaknesses, operational errors and for 'management responses which were lacking in the level of judgement and safety consciousness expected.'

Early in the 1970s, a serious problem had developed which affected the operation of the B205 reprocessing plant at Windscale. In the process of extracting and purifying plutonium, quantities of hot nitric acid were produced containing highly radioactive fission products. This liquid was piped to shielded stainless steel storage tanks in B215 fitted with a cooling system to prevent the contents boiling dry, which would cause serious damage to the tanks and consequent leakage of radioactive matter into the ground. However, in order to reduce the rate at which the number of these costly tanks needed adding to, the cooling system was adjusted to allow some of the liquid to boil away. Nevertheless, with the arrival of spent Magnox fuels the shortage of these tanks became acute, causing a reduction in the reprocessing rate and eventually in September 1972 BNFL had to shut B205 down completely. As a result, spent fuels accumulated in the ponds at reactor stations and at Windscale and this led to their corrosion and deterioration. When B205 started up again, the condition of the spent fuels made reprocessing some-what more tricky and hence slower and so the accumulated backlog was not cleared for several years.

Although BNFL did not disclose these facts, they nevertheless applied for approval to go ahead with what they termed a 'refurbishment' programme at the estimated cost of £245 million. In fact it meant the construction of a complete new spent Magnox fuel processing line and a replacement for B30. BNFL claimed that without it, the continued operation of Britain's Magnox stations could well be affected. When the case came before Cumbria County Council planning committee, outline planning permission was granted forthwith. Interestingly, the Council also granted approval for the development of a process for turning the liquid HLW into a solid, glass-like material. This seemed to be a step in the right direction, faced as BNFL was with the steady accumulation of large volumes of liquid waste.

## The ill-fated Head End Plant

When B205 was commissioned back in the 1960s, the original reprocessing plant B204 became obsolete. So when the UKAEA considered designing a plant to process spent uranium oxide fuels, which would eventually be discharged from Britain's new AGRs and the light water reactors in other countries, it seemed natural to fit it into B204. Hence the latter was gutted and a plant for treating spent oxide fuel, called the Head End Plant, was installed in the building. Here, the fuel was chopped up prior to dissolution in nitric acid and via sundry tanks was pumped over to the nearby B205 reprocessing plant for the extraction of plutonium. The plant was commissioned in 1969 using oxide fuel from the nearby Windscale AGR (see Chapter 13). Technically it worked satisfactorily though it could only operate when B205 was available and in 1972 it suffered about twelve months shut-down.

However when it was time to start up in 1973, unfortunately a build-up of very highly radioactive granules existed on the bottom of a process vessel and by now this was literally scorching hot, unknown to management. When the first quantity of process liquid flowed in, a sudden and violent flashing off of steam took place, causing a blow-out of radioactive vapour into the area where staff were working. Thirty five men suffered skin and lung contamination, mostly from the highly radioactive ruthenium-106 isotope which had escaped. One of these men subsequently died from a high degree of internal radioactive contamination. A report on the accident by the Nuclear Installations Inspectorate (NII) was published in 1974 and it made many recommendations aimed at improving the radiation protection of workers. The main thrust of it concerned the lack of satisfactory emergency arrangements. In fact the Head End Plant never operated again. BNFL wanted a brand new plant with its own reprocessing facilities and a much greater capacity — they were going all-out for foreign processing orders. But planning permission was required and BNFL duly applied to Cumbria County Council.

## THORP

When BNFL's application for its Thermal Oxide Reprocessing Plant, THORP as it became known, was received and discussed at the County planning committee, the latter said it was minded to approve it but that it would involve a departure from the County Structure Plan and would therefore require referral to Peter Shore, the Environment Secretary. On 7 March 1977, Shore announced that there would be a public inquiry into BNFL's proposal for a complete reprocessing plant at Windscale to deal with spent thermal reactor oxide fuel and that it would be held at Whitehaven, a town in Cumbria about six miles from Windscale, commencing on 14 June.

This was something that had never happened before in Britain's nuclear age — a public inquiry into a proposed nuclear investment, and at the Windscale complex too. It was an excellent opportunity for the nuclear lobby and the environmental opposition to argue it out in public, with everything due to be published, in front of Roger Parker, a High Court justice. BNFL must have looked anxiously over its shoulder because of an announcement by outgoing US president, Gerald Ford, to the effect that '*America would no longer regard reprocessing of spent nuclear fuels to produce plutonium as a necessary step in the nuclear fuel cycle.*' Indeed this decision was fully endorsed by incoming president Jimmy Carter in April 1977. He hoped that the American example would help to persuade countries like Britain and France to reject their policies of reprocessing and extracting plutonium. But this went unheeded by James Callaghan, now the Labour Prime Minister.

The Windscale Inquiry into THORP lasted from 14 June to 4 November during which time BNFL and the objectors, the latter consisting of nationwide organisations like Friends of the Earth as well as a number of more local groups, argued the case for and against in great detail. BNFL said the processing of spent oxide fuel was the best way of preparing it for final disposal. But the objectors protested that it was too soon to go ahead since no one knew what arrangements would eventually be made for final disposal, so a case could well be advanced for storing spent oxide fuel intact and all options should therefore be kept open. Furthermore, the chopping up , dissolving in acid and chemical processing of intact fuel turned it into radioactive liquid and gaseous products some of which might well be more difficult to store than the original spent fuel.

Although it was stated that no one had as yet gained experience of successfully operating a commercial oxide reprocessing plant, BNFL said that technically it should be a fairly straightforward business, working forward from their experience already gained in reprocessing at Windscale. On the other hand, it was pointed out, there had been problems at Windscale, in particular with the ill-fated Head End Plant, so how could BNFL be so confident? This was not answered.

From the economic standpoint, BNFL argued strongly that THORP would produce plutonium, and uranium, in suitable form for manufacturing nuclear fuel for which there would be a good market. Objectors thought the cost of extracting and preparing such fuels would quite likely exceed the market price of fresh natural uranium obtained from mining; also, the future need for plutonium for nuclear electricity generation was most uncertain. But both sides at the inquiry were obviously guessing on the topic of fuel prices. And then the argument ran its expected course over the advisability or otherwise of bringing spent nuclear fuel to the UK from around the world; the media used such terms as 'Britain the nuclear dustbin of the World?' However, BNFL argued that it was safer to reprocess for foreign countries at Windscale rather than leave them to build their own plants and thus produce plutonium which could be used in nuclear weapons. But the opposition said they would get the plutonium in any case because the contracts with BNFL would require the latter to return the extracted metal — after all, it belonged to them! The risks of transporting quantities of radioactive materials over the seas were aired but BNFL said they had carefully considered this and balanced it against the advantages of operating what would be a lucrative export business for Britain.

And so it went on, argument and counter-argument, for one hundred days. The objectors, with much less financial resources available for fighting their corner than BNFL, believed that they had made a good case against the proposal. But when the inspector's report was eventually made available to the public, the objectors were dismayed to see that much of their case had been omitted.

Early in 1978, Parker's report was submitted to Shore. Parker had ruled in favour of BNFL and THORP, dismissing all the arguments of the objectors. Although Shore accepted Parker's ruling, the Environment Secretary allowed parliament to debate it, which it duly did in March. On a vote, there were 186 ayes and 56 noes and thus was consent for THORP finally achieved. Those 56 votes must go down as the first formal parliamentary opposition to a major step in the so-called civil nuclear programme. It was inevitable that THORP would be approved, whatever the non-parliamentary opposition in Britain said or did, or whatever the Americans now urged, for there was still a bi-partisan nuclear policy in the House; and one can conjecture that the military lobby would have quietly given support for any processing plant which would produce plutonium. And finally, the jobs question would have come into it, Windscale being the biggest single employer in Cumbria. The opposition at the inquiry never really stood a chance.

But BNFL was now encountering problems in securing contracts for spent oxide reprocessing, both with foreign clients, the CEGB and the South of Scotland Electricity Board. It was apparently in no hurry to go ahead with THORP and did not apply for detailed planning permission for the plant until 1983. Then it had to satisfy the local authority with regard to improved road conditions and a conflict arose over BNFL's wish to extract water for the plant from Wastwater. The organisation was by now already three years behind the schedule advanced at the Windscale Inquiry — it was due to run many more years late before the plant was completed and ready for commissioning.

## Industrial strife at Windscale

Industrial relations had been deteriorating at Windscale for quite a period and it did not require much to trigger an unofficial strike of 3,000 workers in February 1977. Changing-room attendants had been informing management, without gaining a response, that their working conditions were unsatisfactory and so they held a one-day token strike. This resulted in many staff and manual workers being laid off work. The manual workers received less compensation for time loss than the staff and so they stayed on strike. A potentially hazardous situation loomed up since the pickets would not admit carbon dioxide and nitrogen gas supplies into the site; if these gases ran out there could be a risk of explosions. It began to look as though the Energy Secretary, by now Tony Benn, might be forced to call in troops and so he discreetly paid a visit to Cumbria to talk with management, union officials and employees. He heard complaints about 'inept' management, alleged failures to pay compensation for overdoses of radiation and realised that there was a good deal of bitterness amongst the workforce. There was even talk of the Works Manager being removed from his position. However, negotiations eventually led to an end to the strike without troops being called in. Although Benn was not involved in these negotiations, his visit to Cumbria did help to get the two sides listening to each other.

Afterwards, Benn commented to Sir John Hill, chairman of BNFL, that nuclear power had been recommended to him because continuity of electricity supply would not then depend on the goodwill of the miners, and yet look what nearly happened at Windscale. Sir John replied that they would cope by using advanced technology in the design of the next plant. Whereupon Benn queried what they would then do if one of Frank Chapple's electricians was to throw a switch in Carlisle, causing all the Windscale computers to stop; would they be any better off than they were with coal-fired stations dependent on the miners? According to Benn's record, Sir John did not reply.

## The achilles heel

It is clear from the nuclear fuel cycle diagram on page 100 that the radioactive wastes of the nuclear programme finish up at Windscale and we have seen that the most dangerous and longest-lived waste, HLW, emanates from the reprocessing operation on that site. HLW is the achilles heel of the nuclear business. It was not foreseen that it would be so when the nuclear weapon and power programmes got under way.

As we mentioned earlier in this chapter, it was realised at AERE Harwell as far back as the 1950s that one day something would have to be done about the quantities of HLW which would accrue from reprocessing spent fuel from the Magnox reactor programme and hence research and development work continued there until the directive to discontinue it came from UKAEA headquarters.

In September 1976, the distinguished Royal Commission on Environmental Pollution issued its report,: Nuclear Power and the Environment (Cmnd 6618), commonly known as the Flowers Report after its chairman, Sir Brian Flowers, who also happened to be a part-time AEA board member. This report contained the following statement:

'There should be no commitment to a large programme of nuclear fission power until it has been demonstrated beyond reasonable doubt that a method exists to ensure the safe containment of long-lived highly radioactive waste for the indefinite future.'

It also contained a further comment on the matter:

'Neither the UKAEA nor the BNFL in their submissions to us gave any indication that they regarded the search for a means of final disposal of highly-active waste as at all pressing.'

We pick up this all-important subject once more, in the final chapter of this book, but it is noteworthy that Britain continued with the reprocessing of spent reactor fuel at Sellafield, thus separating and accumulating highly radioactive waste, for well into the nineties, whereas Presidents Ford and Carter had declared an end of reprocessing in the US back in 1976. Unless of course they continued with it in secrecy!

## Sea discharges from Sellafield

The start-up of the THORP reprocessing plant on the Sellafield site together with the clearing up of radioactive wastes from earlier nuclear programmes are said to have increased radioactive discharges to the Irish Sea. Nordic countries and Ireland have been demanding an end to these discharges. In addition, the detection in them of radioactive technetium-99 has inflamed opinion in Europe against Britain. In July 1998, Michael Meacher, the Environment Minister, said he was prepared to consider closing Sellafield nuclear plant although this was not the government's preferred option. BNFL has said that although it was working hard to develop technology to reduce radioactive discharges, it would not be possible to stop them altogether.

At the end of July 1998, John Prescott, Deputy Prime Minster, accompanied by Meacher, went to Sintra in Portugal to attend a meeting of members of the Oslo-Paris Convention (Ospar) which controls pollution in the northeast Atlantic. Ospar firmly agreed to reduce radioactive discharges into the sea. This has perhaps signified the beginning of the end for nuclear power stations and nuclear fuel reprocessing plants in Britain. The Sintra agreement is also onerous for the French nuclear industry, which puts radioactive waste into the English Channel thus affecting the Isle of Wight and the Channel Islands.

Prescott is said to have agreed that the eight remaining old Magnox nuclear stations would all be shut down within ten years. The seven, newer, advanced gas-cooled reactor stations and the single pressurised water reactor, Sizewell B, would have to stringently reduce radioactive discharges to the sea. The reprocessing of Magnox fuel is apparently to cease at Sellafield by the year 2020 and if the THORP plant does not considerably reduce its discharges from Sellafield it will also have to close down.

# Chapter 18

# Nuclear power controversies in the seventies

E arly in the 1970s the issue of nuclear electric power generation was becoming increasingly controversial in Britain. Not only were there important differences arising between decision-makers involved with the nuclear business over choice of reactor and the amount of nuclear capacity to be installed, but also there was increasing public concern about the possible effects on the health of employees, and the general public, of pollution from manufacturing industry and the energy supply industries including fossil fuel powered stations and nuclear installations.

Environmental organisations such as Friends of the Earth and Greenpeace were becoming more concerned, particularly about the nuclear cycle. Conditions at Windscale began drawing their attention because (as we saw in Chapter 17) operations and safety procedures at this BNFL site were being criticised by the Nuclear Installations Inspectorate. We also related how the Royal Commission on Environmental Pollution reported to the government in 1976 and was heavily critical of aspects of the nuclear industry, especially the question of reprocessing of spent reactor fuel and the storage of highly radioactive waste.

## Effect of the energy crisis of 1973

Security of energy supplies generally came under scrutiny by government when the Arab-Israeli War of 6-24 October 1973, the so-called Yom Kippur War because of its timing, led to a quadrupling of oil prices by the Organisation of Petroleum Exporting Countries (OPEC). Many members of OPEC were angry at the tacit support given to Israel by the West, although at the outset, Britain had imposed an arms embargo on the Middle East belligerents. In the November, Peter Walker at Trade and Industry announced a ten per cent reduction in all fuel and petrol supplies and the Fuel and Electricity (Control) Bill gave the government power to ration petrol and indeed ration books were prepared, though in the event rationing did not take place.

Early in the month, a miners' overtime ban, a go-slow by power workers and a work-to-rule by rail workers had led to emergency powers being announced by Edward Heath's government. In December he introduced a 3-day week for industry and continuous processes were limited to 65 per cent of normal electricity consumption. In February 1974, the National Union of Mineworkers (NUM) called an all-out strike following an 80 per cent majority in a ballot for industrial action. Heath then called a general election for the 28th. Voting was indecisive, Heath failed to form a coalition and Harold Wilson found himself heading a Labour administration once more, but without a majority. Following the NUM's acceptance of a pay offer, the new government soon announced a return to the 5-day week and also an end to emergency controls on power supplies.

In the January, Heath had already established a Department of Energy and now Wilson continued to develop it, appointing Eric Varley as its secretary of state. As a repercussion from the oil price shock of the previous autumn, and possibly the miners' strike, the government decided to become more informed on energy matters generally and so to work towards establishing an energy policy for Britain. In July 1974, Dr Walter Marshall, then deputy chairman of the UKAEA, was also appointed Chief Scientist to the Department of Energy. This was a newly-created post and though part-time, it carried a good deal of prestige. The choice of Marshall was perhaps unfortunate inasmuch as it gave added weight to the 'nuclear lobby' at Westminster, for he retained his membership of the UKAEA board. It was well known that he was an ardent supporter of more nuclear power and indeed an enthusiastic proponent for building American PWR reactors in Britain.

The Advisory Committee on Research and Development for Fuel and Power (ACORD) was set up by government, drawing its members from the chairmen of the fuel and electricity industries. Marshall was chairman of ACORD and when in 1976 it produced its report, *Energy R & D in the UK*, it was perhaps not surprising that this document accorded a high priority to the fast reactor programme (but see Chapter 14). However, the report also stressed the importance of carrying out energy conservation research and development. Earlier, in December 1974, the government had announced an energy-saving drive aimed at reducing national consumption by 10 per cent.

## Founding of the Energy Technology Support Unit (ETSU)

In April 1974, the Energy Technology Support Unit (ETSU) was established to assess various options open to the UK and to formulate relevant programmes of research and development in the field of non-nuclear energy, especially renewable energy sources such as solar, wind and wave power, and also to examine the scope for energy conservation technology. This Unit was set up in a building on the UKAEA's Harwell site, though it was an integral part of the Department of Energy and had no remit to concern itself with nuclear power. Many of the initial staff of ETSU were ceded from AERE Harwell. Dr Keith Dawson, Head of Applied Chemistry Division at Harwell, was appointed as Head of ETSU. The present writer was recalled from sabbatical as Visiting Professor to Brunel University in order to become Deputy Head of ETSU and to be given special responsibility to build up a team of energy conservation technologists and energy analysts.

The unit's staff, consisting of chemists, physicists and engineers, was built up around two groups; one examined the future possibilities of renewable energy sources such as wind, wave and tidal barrage whilst the other one concentrated on opportunities for introducing and developing energy conservation technologies. Energy auditing techniques were employed for studying the use of energy, industry by industry, and as each one was completed, a report was published jointly by the departments of energy and industry. Thus many instances were revealed where technology to conserve energy existed but needed demonstrating in the workplace. So the audit reports were followed up by the introduction in September 1977 of the Energy Conservation Demonstration Projects Scheme and this ran successfully well into the eighties. The Thatcher administration stressed 'energy efficiency' and indeed the Department of Energy ran a campaign to point out the money-saving advantages of energy conservation in British industry.

Interest in energy analysis methodologies and their applications grew in certain universities, notably Cambridge, Sussex, Southampton, Strathclyde, Sunderland Polytechnic and the Open University. Research contracts were placed with appropriate departments by ETSU. The accounting techniques proved useful for comparative studies between the various renewable energy opportunities and the more conventional fossil-fuel industries. But disagreements arose over the results obtained when the techniques were applied to nuclear electricity generating systems! This stemmed from the heavy capital investment necessary for nuclear generation, for the accounting techniques involved balancing how much energy was consumed in producing electricity with how much was actually generated. It was revealed that much energy was consumed in building nuclear stations, in producing nuclear fuels from uranium ore, and in the subsequent reprocessing or decommissioning of plant.

## Problems with the nuclear power programme

As we entered the 1970s, the AGR stations were giving rise for concern due to technical problems, delays in construction and rising costs (see Chapter 13), although in 1970 the CEGB had ordered its fourth AGR station, for Heysham in Lancashire. The government was being quietly encouraged to develop a new programme against which the construction industry could plan for the future, if indeed it had a future. Also there were varying opinions as to what type of reactor should be selected for the future; the AEA still had a right to advise government on this but the CEGB, the Select Committee on Science and Technology and indeed the construction consortia were increasingly making their views heard.

Edward Heath had set up what was known as the Vinter Committee to help decide what the next choice of reactor should be. This committee did not come up with the answer but it did recommend that the two reactor construction companies should be combined. The government duly accepted this principle; it finally abandoned the pretext of 'competitive' tendering and the precipitating factor was no doubt the shrinking opportunities for the industry. The traditional rate of growth in electricity demand was declining and confirmation of this came from no other person than Arthur Hawkins, chairman of the CEGB. Although he had been thought to be bullish about growth and the need for building more power stations, he now pointed out that annual growth in demand for electricity had shrunk from 5 per cent to 3 per cent and he believed that planning should be based on little more than this figure. That left very little if any scope for more nuclear power after the present programme of AGRs had been completed. In March 1973, Peter Walker at the Department of Trade and Industry duly announced the formation of the National Nuclear Corporation (NNC), to be responsible for all reactor construction in Britain. Later on, a subsidiary of the NNC, the National Power Company (NPC), was set up to do the actual design and construction work. Fifty per cent of the shares in the NNC were to be held by GEC and Sir Arnold Weinstock, chairman of GEC, also became chairman of the NNC.

Now we return to the thorny question of reactor choice. Whilst addressing the Select Committee on Science and Technology, Hawkins had come out in favour of the pressurised water reactor (PWR). However this was not a universal view in the early 1970s. At that time the UKAEA favoured the steam generating heavy water reactor (SGHWR), a 100 MW prototype of which had been performing well at the AEA's Winfrith site in Dorset since 1967 (see Chapter 13). Although Weinstock favoured the PWR, Francis Tombs of the South of Scotland Electricity Board, in his evidence to the select committee, indicated his support for the AEA's SGHWR and so did Lord Hinton. Things looked black for the PWR when the Nuclear Installations Inspectorate (NII) cautioned that it could take a year or two to decide if the PWR, an American reactor, was likely to be licensed in Britain and Sir Alan Cottrell, a well-known metallurgist and Chief Scientific Adviser to the government, was worried about the safety of its steel pressure vessel. Finally, in July 1974, Energy Secretary Varley decided to opt for the SGHWR, one reason being to show the Labour government's support of British enterprise and technology. However since demand for electricity was unlikely to increase very much and installed generating capacity was probably adequate, the government only favoured a 4000 MW programme and even implementation of this was to be deferred for a year or so.

Early in 1975 in a cabinet shuffle, Varley was replaced by Tony Benn as Energy Secretary. Somewhat to his surprise, Benn was informed in July 1976 that the UKAEA had switched its reactor preference to either an American PWR or even the AGR; it had rejected the SGHWR. He consulted the nuclear construction consortium, the NNC, and it transpired that it also favoured the PWR for the future, but as an interim measure recommended continuing with AGRs to keep the industry alive. In January 1978 after more deliberation Benn finally decided to countermand his department's earlier decision to go for SGHWRs and work on that reactor was therefore run down after development costs of nearly £150 million. Instead, Benn authorised the CEGB and the South of Scotland Electricity Board (SSEB) to each order an AGR station. But clearly there was no urgency on the part of either of the boards to take this up. After all, the first AGR, Dungeness B, had still not been completed after twelve years work and the AGR reactors at Hinkley Point B and Hunterston B were yielding less than a third of the electricity they were designed to produce. The options given by the government were not taken up until 1981, when the second AGR station on the Heysham site in Lancashire, Heysham B, was begun by the CEGB, and the SSEB started out on an AGR station at Torness in south-east Scotland. Both these decisions led to vigorous public protests, especially at Torness.

## Serious accident at Three Mile Island

When Benn announced the decisions of 1978 he also added that it was timely to plan for American PWRs for the 1980s. There was undoubtedly pressure building up in the nuclear lobby for the PWR, so that there was almost an inevitability about the PWR crossing the Atlantic, and yet its heyday was over in the US. Orders for nuclear power stations were declining over there in the 1970s mainly for economic reasons; private utilities were facing the true cost of nuclear power while state-owned nuclear monopolies, as in Britain, could hide it — or pass it on to the consumer.

Faith in the safety of nuclear reactors in America took a blow in 1975 when the station at Browns Ferry suffered the most serious safety-related accident to date. An electrician set fire to a cable tray with a lighted candle when carrying out a leakage test. The fire disabled the safety systems on both operating units which were out of action for eighteen months. Nobody appeared to have suffered injury or radiation but public confidence was shaken because of the realisation that carelessness by staff could have such a serious impact at a nuclear plant.

Then on 28 March 1979, a pressurised-water reactor at the Three Mile Island (TMI) nuclear power station in Pennsylvania suffered a more serious accident. For a variety of reasons the feedwater supply system supplying the steam generators of unit 2 failed, causing the reactor core to be partially uncovered. The temperature of the fuel rods rose dramatically and a large hydrogen bubble formed with the serious possibility of a devastating explosion which would scatter radioactivity over a wide region. There was incredible confusion and children and pregnant women were evacuated from a radius of five miles. Radioactivity did escape though the hydrogen explosion was narrowly averted. The affected unit was irreparably damaged and the problems of clean-up were still being grappled with ten years on. Unit 1, although unaffected directly by the accident, only re-entered service seven years after the accident. By order of President Carter, the Kemeny Commission was set up to investigate and report fully on the accident. But long before the report was received, all orders for new nuclear plant in the US were terminated and many years would elapse without any further orders being placed.

It is noteworthy that the impact of the TMI accident was much less in Britain than in the US. It was a PWR reactor and we in Britain were using gas-cooled reactors, which are inherently safer than water-cooled types, as was explained by the South of Scotland Electricity Board chairman at the Sizewell Inquiry (see Chapter 19). One might have expected that the TMI accident would have led to a swing of support in Britain, away from the PWR and back to the AGR. Indeed, the inspector at the Sizewell Inquiry was quite impressed by the presentation on the AGR reactor. But the bandwagon in favour of the PWR was well and truly in motion by the time the Sizewell Inquiry got under way and the massive support for Britain choosing PWR technology ensured that it won the day.

# Chapter 19

# The Sizewell Inquiry

'My conclusions on the risks from low levels of radiation are: It is prudent for the time being to assume that there is no threshold of radiation dose below which there is no effect. All doses, however small, are potentially harmful...'

Sir Frank Layfield, Inspector at the Sizewell B public inquiry, in his official report.

On 3 May 1979, the Conservatives pushed the Callaghan-led Labour administration out of office, winning the general election with an overall majority of 43 seats and Margaret Thatcher became Prime Minister. David Howell became the energy secretary and in a cabinet ministerial committee on 23 October he put forward his proposals for nuclear power.

These were discussed and agreed and at the end of the meeting the Prime Minister summed up as follows. The government's policy would aim to achieve a sizeable nuclear programme. This would have the advantage of removing a substantial portion of electricity production from the dangers of disruption by coal miners or transport workers. The programme would include the prospect of PWRs, subject to satisfactory safety clearances being obtained. The great importance of appropriate presentation for reaching the government's objective was agreed and this meant keeping a low profile approach because opposition to nuclear power might well result in protest groups being active over the next decade. No firm commitment to a PWR would be made until the report of the Three Mile Island accident had been received. The industry would be supported. The Secretary of State was instructed to be guided by this policy.

The plan to adopt the American PWR reactor, with a couple of caveats, might almost be regarded as a continuation of the policy announced by the Labour government's energy secretary in 1978. But the 'low profile' approach to foil anti-nuclear opposition and the plan to increase nuclear electricity in order to become less dependent on the coal miners was a new and unashamed aspect of the Thatcher government's nuclear policy. The CEGB were in agreement with the latter thrust of policy, only they put it more delicately — it was part of increasing fuel diversity.

Aware of growing public concern over safety aspects of nuclear power, the government felt it might have some difficulty in getting its nuclear policy over in the Commons. In December 1979 the Secretary of State for Energy told the House that safe nuclear power and a strong nuclear industry were essential for Britain and the future success of the nuclear programme was vital to the prosperity and security of the country. He said that subject to the necessary consents and safety clearances the PWR should be the next nuclear station to be ordered, with the aim of starting construction in 1982. Furthermore, he envisaged one new nuclear station every year in the following decade. No mention was made about the relationship of capacity to future electricity demand. There was no mention made of how the nuclear building programme might relate to other means of generating electricity — no connections were made with the 1974 Plan for Coal, the feasibility studies taking place on the Severn Barrage electricity scheme or any British-designed nuclear plant such as the AGR. Indeed, the implication was that nuclear power had become synonymous with energy policy. Reference was made by the minister to the fact that there would be need for the fullest explanations and discussions to inform the public inquiry into the first PWR, including presentation of all the principal safety documentation relevant to the gaining of a licence to operate.

The proposed site was not named until October 1980, when it was announced as Sizewell on the Suffolk coast adjacent to the Magnox-type nuclear station. In February 1981 the CEGB formally applied to the secretary of state for consent to build a PWR at Sizewell and a public inquiry was called. However this did not officially start until the beginning of 1983. In the meantime the CEGB and indeed the government had to take some criticism from two quarters.

## Reports in 1981 from the Monopolies and Mergers Commission and the Select Committee on Energy

In May 1980 the government had asked the Monopolies and Mergers Commission to examine the performance of the CEGB with particular reference to its finances and efficiency, and to the effect of these on its customers. The report was published on 20 May of the following year. Its main criticism of the CEGB was with regard to its nuclear investments and its past weaknesses in demand forecasting. However, as regards its investment appraisal method, here there were 'serious weaknesses' evident which led to the board proposing a large programme of investment in nuclear power stations, thus greatly increasing the capital used for a given level of output capacity, and this operated against the public interest.

In 1980, the Select Committee on Energy had begun to examine the new Conservative government's nuclear policy. When it published its report in February 1981 it turned out to be an indictment of this policy. The committee was clearly unconvinced that the CEGB and the government had made out a satisfactory economic and industrial case for a nuclear power programme of the size specified in December 1979. It was critical of CEGB demand forecasting during the course of its investigation, indeed the Board had reduced its peak winter demand for 1986-87 from 52 gigawatts down to 48.5, a 7 per cent cut.

Also in February came a report to the government from the Nuclear Installations Inspectorate entitled *The Management of Safety at Windscale*. As we have already recorded in Chapter 17 this was very critical of BNFL with regard to conditions at the Windscale site. The report had been long delayed in coming, it is thought partly due to staff shortages and partly because of the amount of work which the NII had found it necessary to do at the site. When it did come, it was an awkward time for the government.

Indeed, these three reports, coming as they did shortly before the expected start of the Sizewell Inquiry, critical as they were of CEGB, BNFL and the government itself, must have worried the proposers of the PWR and at the same time heartened some of the potential inquirers! The latter would also have been pleased in July 1981 when the secretary of state said that the inquiry would have to take into account the need for a secure and economic electricity supply, bearing in mind the government's long-term energy policy; this, they believed, would allow testing of the government on long-term energy issues and thus the inquiry would not be merely a local planning matter. At this time, the government published a short White Paper, rebuffing the select committee's recent report and reaffirming its concept of a programme of nuclear plant building. But it was having enough trouble over selecting the design of its first PWR. This was causing delays in having a reference design available in time for the NNI to be able to make a safety assessment of it in time for the inquiry in 1982. So it was perhaps no real surprise when Nigel Lawson, now Secretary of State for Energy, announced in January 1982 that the inquiry into Sizewell B PWR would open a year late, in January 1983.

## The Sizewell B Inquiry opens

The Sizewell B Inquiry duly opened at Snape Maltings in Suffolk on 11 January 1983 under its Inspector, Sir Frank Layfield, QC. It closed on 7 March 1985, having run for 340 working days.

It was always expected to be a costly affair but no one anticipated it would last as long as it did. The CEGB alone is thought to have spent about £20 million over it. Protesters were worried about their likely costs and the inspector held three pre-inquiry meetings between June and October 1982 when he was pressed to try and arrange funding from the government for bona fide objectors. He failed to achieve success with the secretary of state who rejected his request essentially on grounds of precedent. So objectors had to face paying for their own legal representation, the expenses of any expert witnesses and the carrying out of supporting research. After the final decision was made regarding funding, only two voluntary organisations chose to retain legal counsel, the rest deciding to do the job for themselves. This had to be balanced against the CEGB's team of four barristers and over 25 support staff.

There were about 25 objecting organisations, ranging from the larger bodies such as Campaign for Nuclear Disarmament (CND), Council for the Protection of Rural England (CPRE), Electricity Consumers Council (ECC), Friends of the Earth (FOE), The Greater London Council (GLC), National Union of Mineworkers (NUM), Town and Country Planning Association (TCPA) to small groups such as the Suffolk Preservation Society (SPS), Stop Sizewell B Association (SSBA) and Billingham Against Nuclear Dumping (BAND). There were some objecting organisations who did not take part in the inquiry, not necessarily for lack of funding, but rather because they thought the outcome was a foregone conclusion and the whole business was simply a public relations exercise on behalf of a government which knew what it wanted and was determined to get it. They may have been right, but many matters relating to nuclear energy and its politics, important matters, were aired publicly which might not have been otherwise. It caught the attention of the media, books were written about it. But there is no space here for a systematic account, or even a summary, of what was discussed throughout those 340 working days.

Apart from matters of particular importance to people living and working in the area of Sizewell and the town of Leiston, there were three, more generic topics dealt with at some length, all of key concern regarding the future of nuclear energy, if it was to have any long-term application other than a purely military one:

(1) the economic and strategic case for the CEGB building a 1200 MW PWR station;

(2) safety aspects of the design of this reactor;

(3) the problem of dealing with radioactive waste.

We now deal briefly with each of these items in turn.

# 1   The economic and strategic case for the PWR

The CEGB central forecasts for required electricity generation showed no shortfall before 1997, so the economic case for Sizewell B had to depend on the electricity generated throughout its lifetime being less costly than that from any other source. The savings would have to be achieved by having coal and oil-fired generating plants shut down for long periods or altogether. Whether or not this was likely depended upon the estimated future price of these fuels. There was much argument at the inquiry over the likely figures; clearly no one could know them, since the markets for these fuels can be very volatile. By the time the economic arguments had finished, many participants considered that the economic benefits would be less than the CEGB had originally claimed. Indeed, the ECC had concluded that the economic case for Sizewell B was 'marginal verging towards adverse,' and that consumers could not benefit to any large or direct extent by constructing the station ahead of capacity need. A result of this trend at the inquiry was that the CEGB began to lay more emphasis on the strategic argument in supporting its case for Sizewell B, essentially that the station would reduce the preponderance of coal generation in the electricity supply system. It would increase fuel diversity and hence security of electricity supply. But against this it was argued that the effect would only become significant if a series of nuclear stations were built; the case for Sizewell B was supposed to be convincing whether or not further PWRs were built. In any case, increasing dependence on an imported fuel (uranium ore) would be regarded in many quarters as a peculiar policy for gaining increased security of electric power supply.

Early in the inquiry it was unclear whether there would be an ongoing programme of PWR stations. The CEGB had talked about it making sense to have a small group of PWRs rather than one, indeed it had announced its intention to gain agreement for a further station at Hinkley Point adjacent to existing nuclear reactors. Towards the end of the inquiry, counsel for the ECC declared that evidently the intention was to build a series of PWR stations rather than just one, although the government's actual intentions were never really clarified.

# 2   Safety aspects of the design

Since the proposed PWR for Sizewell B was based on a design entirely new to Britain, it was to be expected that the NII would need a lengthier time than usual in which to decide whether it conformed to UK safety standards. Until a positive conclusion was arrived at, no licence for construction could legally be issued. Since the NII were still at work on their assessment of Sizewell B during the inquiry, the CEGB could not know whether or not they would be allowed to build the station, irrespective of the ultimate outcome of the inquiry. The counsel for FOE applied to the inspector for an adjournment of those parts of the inquiry concerned with safety but after considering this for several months he rejected it on 14 June 1983.

The government had earlier said that the NII would complete its work in time for the opening of the inquiry. Since this was going to be impossible, the only solution would have been, once again, to postpone the start of the inquiry. The secretary of state and the CEGB did not want this to happen. It may be fairly asked, was it the NII's fault that they could not finish on time? No, it was probably because the CEGB changed the design of the reactor after the NII had been working on it for some months and the chief reasons why the Board changed the design were because it would have been too expensive, would have taken too long to build and did not make the best use of American experience. The new design was produced by a so-called task force appointed by the secretary of state and led by Dr Walter Marshall who was at that time chairman of the UKAEA but was shortly to be appointed to take over at the CEGB. His team included senior engineers from Bechtel, an American firm of construction engineers, and also from Westinghouse, the American company responsible for the initial design adopted by the CEGB. Although the new design was fairly quickly prepared, it nevertheless was going to take longer for the NII to carry out their task because of the change from what was a British design to an essentially American one. So in the end, the Sizewell Inquiry and the licensing inquiry had to proceed at the same time.

When the PWR Project Director was cross-examined by Friends of the Earth over the relative safety of the two designs, the main difference seemed to lie in the fact that in the first design, the four sets of key safety machinery were equally spaced in an annular building around the reactor containment dome whereas in the second design the sets were grouped together on one side of the dome which reduced their independence and segregation, resulting in a diminished resilience to common cause failure. However, the latter design would be cheaper and quicker to build. In short, the earlier design would, on balance, be safer but the second design would be cheaper!

There was much discussion at the inquiry about the intrinsic safety of an AGR reactor compared with a PWR. The gas-cooled AGR would have the facility of maintaining itself sufficiently cool by natural convection of carbon dioxide gas in the event of coolant circulation pumps failing. But in the case of such a failure with a PWR, the static water coolant would be likely to turn to steam which would not prevent reactor core temperatures rising, with resultant melting of the fuel cladding and perhaps the fuel itself, as happened in the Three Mile Island accident. The inspector's interest seemed roused and he invited the South of Scotland Electricity Board chairman, Mr Donald Miller, to give evidence. There was no doubt about his support for the AGR. His evidence also suggested that the AGR would be very close to the PWR in economic terms.

## 3 The problems of radioactive waste

An examination of the diagram of Britain's nuclear fuel cycle on page 100 will reveal the connection between CEGB nuclear reactors and the various radioactive wastes; any increase in nuclear capacity will inevitably lead to an increase in the amount of wastes. Therefore the protestors at the Sizewell Inquiry naturally objected to the addition of a 1200 MW nuclear generator to the Board's capacity, maintaining that wastes would tend to escape into the environment and hence the risk of cancer cases would increase at Sizewell and the surrounding area. Also, the amount of radioactivity would increase at Sellafield if and when spent fuel arrived there from Sizewell B.

Public concern over nuclear waste increased generally after November 1983 when accidental discharges of radioactive materials into the Irish Sea took place during annual maintenance work at Sellafield. Quite by chance Greenpeace encountered one of the discharges whilst sailing off the pipeline outlet. BNFL was duly prosecuted. This event occurred only a few days after the television documentary film *Windscale: The Nuclear Laundry* had been shown. Its main focus was on the incidence of cancer in young people resident in the area of Cumbria near to the processing site. The leukaemia rate for children was particularly high at about ten times the national average. The film presented evidence of radioactive material actually being washed back onto beaches, which had to be closed off for some time. These so-called 'leukaemia clusters' were also located in other areas of the country.

The East Suffolk District Medical Officer, Dr Michael Bush, carried out a study of leukaemia incidence in his area after public interest had been raised by press reports of cases being found near Sizewell A. He was invited by the inspector at the inquiry to report his findings. Under cross-examination, he agreed that there seemed to be a greater susceptibility to leukaemia among the people of Leiston (the nearest town to Sizewell, population about 5,000) than is normal. He also said that research was taking place into the question of the levels of radiation which may or may not be leukaemogenic and it would be preferable to have the results of these studies before making a decision to increase possible radiation levels within a particular locality. This was taken by the Sizewell objectors as a strong argument in their favour.

There was a lot more discussion of health and safety matters regarding Sizewell B at the inquiry but space precludes the covering of any more evidence or discussion of this topic area.

The inquiry closed on 7 March 1985 but the inspector's report was not published until January 1987, nearly two years later. In the meantime, a national coal strike led by the National Union of Mineworkers took place in 1984 and a serious nuclear explosion occurred at the Chernobyl nuclear power station near Kiev in the Ukraine. Both these events and their likely impact on the minds of the public in Britain — and the government — must have given the inspector much food for thought.

## January 1987 — Report on Sizewell B is published

On 13 November 1986, Roger Milne, writing in *New Scientist*, stated that Peter Walker, now Secretary of State for Energy, was about to receive Sir Frank Layfield's 100-volume report into the two-year Sizewell Inquiry. The outcome was evidently a foregone conclusion so far as the government was concerned. 'A decision by the cabinet to sanction the £1.3 billion project is expected in the new year, once the report is published,' said Milne. However, the NII had not yet granted a full safety clearance. A number of safety issues about design still remained. One concerned the reactor's coolant pumps and another was provision for the inspection of both pressurisers and the steam generator, key components of what is called the primary circuit. Also, since the inquiry closed, the CEGB had admitted that it would have to spend £8 million on building a concrete barrier underneath Sizewell B to protect the neighbouring Sizewell A station from the effects of any fall in water table due to building Sizewell B.

In January 1987, the report was finally published and *The Guardian* heralded it with the following headline:

LAYFIELD OPTS FOR BUILDING POWER STATION DESPITE HEALTH AND ECONOMIC RESERVATIONS

This said it all, a wonderfully brief and accurate statement of the inspector's lengthy conclusions.

In his general conclusions he stated that the disadvantages of risks to health and safety and of environmental damage to the Sizewell locality must be outweighed by anticipated benefits for the nation. There was a national interest in building a PWR. That national interest could best be met by building it at Sizewell. He said the CEGB had discharged the onus of proof that national need should override the local interest in favour of conservation. 'In my judgement the expected national economic benefits are sufficient to justify the risks that would be incurred.' So-called 'national need' was well plugged in Layfield's report, so he would be regarded as having done a sound job for H. M. Government, just as Roger Parker, the High Court judge, had done in the case of the THORP Inquiry (see Chapter 17).

The inspector at Sizewell clearly had not brushed aside all concern about the possible relationships between cancer and various levels of radiation. He was cautious. For example he called for data on leukaemia in all workers at Sizewell A to be collected and analysed and indeed assessed in conjunction with similar data on all present and former CEGB nuclear power stations. Studies were needed on more general patterns of cancer incidence among the public; he stressed that the public would be exposed to radiation even as a result of Sizewell B operating normally. The inspector was quoted at the head of this chapter as having written that he felt it prudent to assume, for the time being at least, that there is no safe threshold of radiation dose below which there is no effect.

The government soon gave authorisation for the CEGB to go ahead and build Sizewell B when its licence was forthcoming from the NII. Construction was under way by the end of the year.

However, a major change in government policy on nuclear power was made in November 1989 and as a result, plans to build any further PWR stations beyond Sizewell B were cancelled. But construction of the lone PWR continued and Sizewell B duly opened in February 1995. As of January 1999, no further applications for building nuclear power stations in Britain have been received.

*Note*

In drafting this chapter, a good deal of factual information about the Sizewell Inquiry was taken from *Atomic Crossroads: Before and After Sizewell*, Merlin Press, 1985. It was written by John Valentine, to whom the present writer is much indebted.

# Chapter 20

# Privatisation and nuclear power

Toward the end of 1987, the government announced that it had decided to reconstruct the UK electricity supply industry and Cecil Parkinson, the energy secretary, put out the broad policies for doing this in the form of a White Paper. Britoil had been sold off in 1982, British Gas in 1986, and now it was the turn of electricity to be privatised. It was not going to be easy, said Parkinson, for some twenty-four acts of parliament governing the industry and stretching back for over a century would have to be recodified and included in the Bill which he hoped to present by the end of the following year. In fact it took much longer. The tactic was to make the industry appear consumer-led rather than producer-dominated. Competition would be introduced into the industry in place of the present monopoly situation. Increased efficiency would result and lower electricity prices would follow.

But the CEGB, led by its chairman, Lord Marshall of Goring, maintained that its record of a consistent electricity supply to the nation was near faultless. He accepted that electricity supply was a natural monopoly and he was all for retaining the National Grid and most of the generating capacity as a single command structure, this being the best means of holding down electricity costs. Marshall was on his own against Parkinson and heavyweights like Chancellor Nigel Lawson and Trade and Industry Secretary Lord Young, despite the fact that Lawson had asked him to become chairman of the CEGB back in 1982. The cabinet under Thatcher wanted control of electricity to rest with those selling it directly to the public, namely the twelve area boards, which took their supplies from the grid. Marshall lost his case. Furthermore, he also found he was running against the government when he expressed doubts over whether it would be possible to privatise the nuclear power stations, given their financial liabilities for fuel reprocessing, decommissioning and waste storage. The House of Commons Energy Committee estimated that the decommissioning of each nuclear station would cost somewhere between £250 and £750 million.

In 1989, the City had the final say in the economic debate over nuclear electricity costs, *maintaining that nuclear power was not economic and thus firmly rejecting any plans the government might still have in mind for including part of the nuclear industry in the privatised electricity industry*. It was an embarrassing climbdown for the government, which was forced to revise its plans for privatising electricity and had to retain the nuclear stations in the public sector.

## The new Electricity Act, November 1989

The Act provided for a restructuring of the electricity industry so that its assets could be vested in a number of new companies. Vesting day was 31 March 1990, the date on which the Act came into force, after which the companies were prepared for floating on the stock market. The Act only covered England, Wales and Scotland. Northern Ireland was dealt with later under the Electricity (Northern Ireland) Order of 1992.

Prior to privatisation, generation and transmission of electricity in England and Wales were the responsibility of the CEGB, the twelve area electricity boards being responsible for distribution and supply. After privatisation, the CEGB's generation activities were shared among three companies. National Power and Powergen were the two non-nuclear parts, floated on the stock market in March 1991, with the government retaining 40 per cent of the stock — although all or most of these 'golden shares' were intended to be sold in February 1995. Nuclear generation, intended originally to form part of National Power, was transferred to a third company, Nuclear Electric, which remained in public ownership. Marshall had been offered the chairmanship of National Power but for him it was the end of the road and he resigned. Becoming the chairman of the CEGB had marked the zenith of his public service career. He had been a powerful figure in the 'nuclear lobby' for many years; his going signalled the beginning of the end for that lobby.

The CEGB's transmission activity, namely the National Grid, was transferred to the National Grid Company (NGC), which also took on board the CEGB's pumped-storage power generation stations.

In Scotland, prior to privatisation, joint responsibility for generation and transmission activities were shared between the North of Scotland Hydro-Electric Board (NSHEB) and the South of Scotland Electricity Board (SSEB). After privatisation, the SSEB became Scottish Power and the NSHEB became Scottish Hydro-Electric and their shares were sold on the London Stock Exchange in June 1991. However, the SSEB's nuclear generation activities were transferred to a new company, Scottish Nuclear, which remained in state ownership.

The twelve area boards were given independence and floated on the stock market in December 1990 and were known as the Regional Electricity Companies (RECs). Although their core business was distribution and supply, most of them also began to diversify into generation — gas-fired power stations had become popular. It was the time of the so-called 'dash for gas'.

The Regional Electricity Companies owned the National Grid Company through a joint holding company although the latter was floated on the stock market in 1992. For this purpose, it was only valued by the government at £1 billion but five years later, its stock market valuation was more like £5 billion. A crucial introduction at privatisation was the setting up of a wholesale market for electricity known as the Electricity Pool, which acted as a trading interface between generation and supply.

How all these complex changes have worked out for the consumer in terms of electricity charges as compared with the old days when electricity was a public service is a thorny topic outside the scope of the present book, for we are only concerned here with the nuclear aspect of supply and its effect on charges. But an important inflating effect on electricity prices has been the 10 per cent levy imposed by government, known as the Non-fossil Fuel Levy, which the privatised industry had to add onto their bills. The intention of this was to provide finance for the development of non-fossil fuel energy supplies such as wind and wave power and also to offset the costs of decommissioning nuclear power stations. As stated earlier, the House of Commons Energy Committee had estimated a figure for this of between £250 and £750 million per power station. In the event it is generally understood that a good proportion of the levy went to subsidise nuclear electricity generation instead of going towards decommissioning costs.

## The 1995 government review of nuclear power

At the time of privatisation in 1990, the government ordered a five-year halt on the building of any new nuclear stations after Sizewell B, pending a review of the longer term outlook for the industry. The review was duly published in May 1995 and it showed that despite gains in efficiency, nuclear power was still hopelessly uncompetitive. Furthermore the review demolished, one by one, the nuclear industry's arguments for special treatment by government, so there would definitely be no cash made available from that quarter. Ministers effectively ruled out the building of any new reactors in Britain for the foreseeable future. This policy was not running against general world trend. No new reactor had been ordered in the United States for seventeen years and nuclear expansion had been halted everywhere in Western

Europe except France. Scores of planned reactors had been cancelled in the former Soviet bloc.

Probably the most surprising recommendation of the nuclear power review was that the electricity generating section of the nuclear industry was to be privatised — or rather part of it was. Nuclear Electric (in England and Wales) and Scottish Nuclear would be combined and then split into two new companies. The nine oldest of the Magnox power stations, three of which had already closed down, were to remain in a publicly-owned company which would handle the gradual closure of all the Magnox stations. It was intended that the stations would eventually be handed over to the state-owned British Nuclear Fuels although John Guinness, chairman of BNFL, demanded that the government address the issue of decommissioning and give financial guarantees before his company accepted that it should take over the ageing plants.

The newer nuclear stations — seven AGRs plus the recently-opened Sizewell B PWR station — would be sold off to private investors as a single package. The government floated this new company on the stock exchange in July 1996. However, persuading private investors to buy it cost the government a good deal of money, said to total £1 billion over three years following the decision to sell off the modern reactors. The Labour opposition pressed the government to halt its nuclear privatisation plans but this call was ignored. The new nuclear company was called British Energy — the omission of the word nuclear from the title seemed significant — and indeed it contemplated diversifying into gas-fired generation for there was no legal reason against doing this. But plans for such a plant at its Heysham site in Lancashire were abandoned when the company's performance did not please the City. British Energy announced the shedding of 1,460 jobs in the autumn of 1996 raising questions over whether the remaining labour force could be relied upon to maintain high standards of safety. The proposal for building another PWR at Sizewell had been abandoned since it would possibly have required a subsidy of at least £1.4 billion and the government maintained that there was no justification for this.

## The privatisation of AEA Technology

AEA Technology was the science and engineering arm of the UKAEA which began to develop following the passing of the Science and Technology Act of 1965. Section 4 of the latter allowed the UKAEA research and development laboratories to become involved in non-nuclear joint projects with industry (see Chapter 15). During 1995-96, preparations were in hand for privatising this successful company and floating it on the stock exchange.

Government advisers recommended that AEA Technology be separated from the UKAEA as the latter was considered unsaleable because of its responsibilities for contaminated sites and the decommissioning of a variety of shut-down reactors. The complexity of the separation process cost a lot more than had been anticipated due to the intermeshing of laboratories, plant and supporting facilities. A costs figure of over £100 million was finally reached, this at a time when the company was only expected to fetch around £200 million at its sale. AEA Technology would still be located in laboratories on land belonging to the rump of the UKAEA to which it would pay rent. In return, the UKAEA would hire scientists and engineers from AEA Technology.

## No privatisation of BNFL

The government's plans for the privatisation of the nuclear business excluded British Nuclear Fuels Ltd. As mentioned in Chapter 15, the reorganisation of the UKAEA in 1971 had resulted in its Production Group being hived off to form the state-owned company of BNFL.

During the NPT Conference in 1995 (see Chapter 24) Britain and the USA were criticised by non-nuclear weapons states for failing to make progress towards nuclear disarmament. In response, both countries stressed that they had ceased to produce plutonium for weapons. They did not mention that they had large stockpiles of this element but intended to go on making tritium because it has a half-life of only about twelve years and therefore has to be replaced periodically. Because the continued production of tritium could promote the spread of nuclear weapons, some scientists have called for tough controls on its export. At the NPT Conference, a group of eleven countries including Australia and Canada called for consultations to try and control tritium exports.

So we must conclude that BNFL was rejected for privatisation because of its involvement with the production of nuclear weapons materials, which required very stringent security controls.

Part 3

# BRITAIN'S CONTINUED INVOLVEMENT WITH NUCLEAR WEAPONS

# Chapter 21

# Attempts to control nuclear weapons

W e have already seen (Chapter 8) that since 1953 the testing of nuclear weapons had been increasing the levels of radioactivity in the world's atmosphere. Britain had commenced regular testing of nuclear fission bombs and the US, having developed the H-bomb, was testing them regularly. The Soviet Union also developed the H-bomb and carried out test explosions on a fairly regular basis. Albeit on a smaller scale, accidents associated with nuclear weapons and their manufacture had added to this increase in aerial radioactivity. There was also the ever-present risk of a serious accident taking place. Indeed, there was a huge accidental release of radioactivity in the Urals (Soviet Union) in 1957, although this was not known in Britain until 1976.

## The need for a test ban treaty

When world leaders met in Geneva in 1955 there had seemed to be the possibility of a *rapprochement* over nuclear weapons (Chapter 7) but nothing came of this. Growing concern in Britain about the health-related risks from nuclear power and weapons programmes led to the formation and growth of CND in the late fifties. Even amongst those who believed Britain should continue to have its own independent nuclear deterrent, quite a few were against nuclear weapons testing.

In those days there was no means of testing weapons in the laboratory by simulation techniques because computer technology was in its infancy so a comprehensive test ban, if honoured, would virtually have brought the development of nuclear weapons to a halt. Some discussions about a test ban treaty did take place in the UN, indeed an informal moratorium on atmospheric testing had been observed since 1958, but there was no real will on the part of the superpowers to come to any formal agreement until 1963, in the aftermath of the unfortunate crises now to be described.

## The Berlin Crisis

In the first half of 1961, President John F. Kennedy's special disarmament adviser John J. McCloy had been meeting with Soviet Deputy Foreign Minister Zorin to discuss procedural matters relating to possible resumption of negotiations on a test ban treaty, but these ultimately foundered on the perennial problems of verification and frequency of site inspection. In any event, the discussions were overshadowed by the Berlin Crisis which dominated the summer months of 1961. The East German government was concerned about the number of refugees moving to the West and thus reducing its labour force. The Soviets had taken up their cause and threatened to end Western access to the city. Kennedy responded in July by announcing additional defence procurements to the tune of $3.25 billion, including more inter-continental nuclear-warheaded missiles. During the night of 17 August the first sections of the Berlin Wall appeared. Then the USSR decided to break the moratorium on weapons testing in response to the West's reinforcement of the Berlin garrison and more widespread troop mobilisations.

During the autumn of 1961, the Soviets held a series of about 50 nuclear explosions, the largest having a force of around 57 million tons (megatons). In response, the American chiefs of staff pressed Kennedy to start testing again and this began with underground explosions in September followed by atmospheric tests in April 1962. In an endeavour to show that it was still one of the Great Powers, Britain itself began testing in March 1962.

There was world-wide revulsion at this resumption of nuclear testing and fears grew about the radioactive fall-out which would result from it. There were renewed discussions in the UN General Assembly, starting in March 1962, about a comprehensive test ban treaty for nuclear weapons. During the summer it did look as though the US and the USSR were serious about achieving an agreement at Geneva. But there were still verification problems to be resolved when the autumn came and along with it a crisis between the two superpowers which might easily have led to a major war.

## The Cuban missile crisis

In the spring of 1961, a force of Cuban exiles, trained and equipped by the Central Intelligence Agency (CIA) in America, had landed on Cuba at the Bay of Pigs in an attempt to overthrow the communist regime of Fidel Castro, who had gained power in January 1959. The invasion went badly wrong. Khrushchev roundly condemned the American government and was apprehensive about possible further attempts to overthrow his client, Castro.

On 16 October 1962, President Kennedy learned that Soviet technicians were engaged in installing medium-range ballistic missiles (SS-4s and SS-5s) in western Cuba. Since these nuclear weapons, virtually on America's doorstep, could strike the US within minutes of launching, the president decided that they had to be removed forthwith and by the 22nd, the armed forces were on alert. A naval blockade of Cuba then took place, 200,000 troops were assembled in Florida and tactical fighters moved to within striking distance of Cuba. Work still continued on the missile emplacements for a while but were halted on 28 October when Kennedy and Khrushchev were able to come to an accommodation.

Khrushchev had consistently asserted that the rockets were intended solely for the defence of Cuba against any further invasion. When Kennedy gave him firm assurances on that score, he promised 'in return' to withdraw the Soviet weapons. But there was probably another dimension to the situation; Lord Healey later pointed out that Khrushchev had not reacted when the US had put ballistic missiles into Turkey, only a few hundred miles from Soviet territory and he therefore thought it unlikely that America would react if he put missiles into Cuba. Indeed, when Khrushchev took them out, Kennedy removed the American missiles from Turkey.

## The partial test ban treaty

Although the superpowers appear to have had insufficient will to conclude a test ban agreement prior to the Cuban crisis, the shock of the confrontation now seemed to generate a will to get something done. The stumbling block of the Soviets' inability to agree on the matter of on-site suspect seismic tremors inspection seemed to be cleared when Khrushchev announced that he would agree to two or three per year as well as to the siting of three automatic recording stations in seismic zones of the Soviet Union. But it was not enough for the USA which continued to insist on a minimum of eight to ten inspections. Also, the proposed locations for the unmanned sensors were thought to be unacceptable. Neither side would move to close the gap and so the possibility of a complete nuclear test ban treaty was being allowed to slip slowly away.

However, Kennedy made an important speech on 10 June 1963 at the American University, and this did seem conciliatory. He plugged the line that total war made no sense and he called for an examination of the Cold War mentality by many Americans. At the end of this speech, the president announced that high-level negotiations would shortly begin in Moscow to try to break the deadlock over the test ban treaty. The president's speech was received with enthusiasm in the USSR and indeed given extensive news coverage. It was praised by Khrushchev on 2 July 1963 who then offered a limited test ban to be agreed between the superpowers. Within ten days, 15-25 July, the test ban was resurrected, agreed and signed.

Britain succeeded in getting herself included in these negotiations by virtue of her independent nuclear capability although she contributed nothing of a positive nature, indeed, she was simply 'playing the self-important retainer to an indulgent American potentate to satisfy the delusions of grandeur still entertained by sections of the British political community' (to quote from page 121 of *Defended to Death*, edited by Gwyn Prins, Penguin Books, 1983). Britain did, however, succeed in weakening the draft of the partial test ban treaty by getting an escape clause inserted into it (Clause 4), permitting abrogation of even its limited provisions if the 'supreme interest' of a member state was felt to be jeopardised.

There was no timetable set for progressing towards a *comprehensive* test ban treaty although the abrupt switch from comprehensive to partial agreement was seen at the time as only temporary, a concession to political expediency. Nothing in the treaty prevented the development of nuclear weapons by the use of underground testing methods, as indeed was the practice to be followed for decades. However the treaty was welcomed as a public health measure because it reduced the rate of increase of atmospheric radioactive fall-out, though it gave the erroneous impression that action by governments to control the arms race was in hand. Indeed, membership of CND in Britain fell away to some extent after its introduction. Protest against nuclear arms also declined in other countries.

## The Undén plan

In 1962 there had been a move at the UN to prevent the further proliferation of nuclear weapons. This centred on a proposal from the Swedish Prime Minister, Osten Undén, to create a non-nuclear club of nations who would undertake not to manufacture or obtain nuclear weapons or permit to be deployed on their territory those of any other power. The idea was to combine the control of nuclear proliferation with the beginnings of positive movement from a regional nuclear disarmament to general nuclear disarmament. The Undén scheme was supported by the Soviet Union, the Warsaw Pact countries, most of the non-aligned nations and four members of NATO (Canada, Denmark, Norway and Iceland), but was opposed by twelve nations, ten of which were the remaining NATO members including of course Britain. But the scheme did not blossom because it was overtaken by the partial test ban treaty in the following year.

### The non-proliferation treaty

Although the Undén scheme did not of itself take off after 1962, nevertheless the idea simmered on. Many non-nuclear states (NNS) had witnessed the growth of military power achieved by virtue of possessing a range of nuclear weapons and simultaneously the nuclear weapon states (NWS) had become increasingly conscious of the dangers which could stem from other states acquiring these weapons. In the 1960s, only five states admitted to the possession of nuclear weapons — the US, UK, USSR, France and China. They also happened to be the five permanent members of the United Nations Security Council! By 1968, they saw to it that the UN had drafted a non-proliferation treaty (NPT) which prevented the acquisition of nuclear weapons by those states which did not already possess them. It was to be effective for 25 years after which it could be extended. The Treaty defined different conditions as applying to the NWS and NNS and it was to come into force in 1970.

Of course the NNS could not be compelled to sign the treaty, so there had to be some inducements coming from the NWS in order to get the NNS to pledge themselves not to acquire nuclear weapons. So the NWS on their part pledged themselves:

(a) not to transfer nuclear weapons to other states;

(b) not to assist NNS to acquire nuclear explosive devices; and

(c) to negotiate in good faith the achievement of nuclear disarmament throughout the world.

In signing the treaty it became mandatory for the NNS to accept inspection by the International Atomic Energy Agency (IAEA) of any nuclear materials or plants under their jurisdiction in return for the freedom of the said NNS to develop civil nuclear energy for peaceful purposes only. This was a weak point in the treaty for it was giving the IAEA a difficult task — because as we have seen in Chapters 11 and 12, it can be very difficult to identify the difference between so-called civil and military plants and materials. And of course, if a NNS refused the IAEA permission to enter a building to inspect, there was nothing the latter could reasonably do about it except to go home and complain to the UN. The IAEA do also inspect nuclear plants in NWS countries but they must accept the word of the host country if they are refused entry on military grounds. Indeed, it is believed that most of Windscale site is barred to the IAEA so that no inventory of British fissile material stocks can be made.

Although Britain signed the non-proliferation treaty by 1970 when it came into force, along with the US and the Soviet Union, the remaining NWS, France and China, did not sign for many more years. The number of NNS signing steadily increased until by the time of the treaty's review in 1995 it had reached 167. Among the more significant non-parties to the treaty in 1970 were Israel, India and Pakistan, who were generally believed to be in the process of acquiring nuclear weapons or, at the very least, putting themselves into a position to be able to construct such a weapon at very short notice. These countries have been termed 'threshold' nuclear powers. Although Iran, Iraq and North Korea eventually signed, they have sometimes been referred to as threshold powers because they were nevertheless believed to harbour nuclear intentions.

The threshold powers and indeed many NNS took the view that the treaty was a discriminatory imposition of the 1970 nuclear status quo on the rest of the world. They were cynical about when and how far the NWS would go to fulfil their part of the bargain. Indeed, it has to be said that it is difficult to discern just what Britain did about 'negotiating in good faith on nuclear disarmament' (Article VI of the Non-Proliferation Treaty). Also, it was known that the NWS were steadily increasing their nuclear weapon capacity in 1970; Britain was just commissioning its Polaris submarine force and was shortly to introduce the Chevaline programme. How could this be said to be within the spirit of 'non-proliferation'? But further disarmament measures were discussed at the United Nations after 1970, indeed several more treaties were drafted and we now briefly refer to them.

## Other nuclear treaties

Back in 1967, the Treaty of Tlateloko was signed, prohibiting nuclear weapons from entering Latin America, although Argentina and Brazil eventually became involved in thinly-disguised nuclear weapon research and development. Two partial treaties were drafted; there was the Threshold Test Ban Treaty of 1974 which limited the yield of underground tests to a force of 150 kilotons, and then in 1976 a Treaty on Underground Explosions for Peaceful Purposes was drafted. But neither was ratified.

A number of disarmament treaties tended to be agreements not to do things which it was unlikely that anybody would wish to do. But the mere fact of passing them and publishing the fact would tend to delude the public in general into thinking that the UN was steadily working towards nuclear disarmament. One example was the Sea Bed Treaty of 1972 which prevents the placing of nuclear weapons on the sea-bed. Then there have been agreements not to do what could not be done at the time the agreement was made. Perhaps the main example was the Outer Space Treaty of 1967 which was intended to prevent space from ever becoming militarised. But later the treaty was breached under a 'civilian' disguise: the Americans with the Space Shuttle and space-based laser weapons, and the Soviets with anti-satellite weapons.

Then there are the SALT and ABM treaties which were really agreements between the US and the USSR and hardly involved Britain. The SALT I negotiations, strategic arms limitation talks, produced a treaty which restricted each side to two sites for the deployment of anti-ballistic missile (ABM) systems and this was actually put into force on 3 October 1972. SALT I also produced an interim agreement for limiting the numbers of launchers and delivery systems and the aim was to lead on to a comprehensive treaty on offensive weapons at SALT II. The Cold War came to an end and the Soviet Union broke up before this was reached.

*Note*

When India carried out five nuclear explosions under the Rajasthan Desert in May 1998, people in Britain were duly shocked as indeed they were when, fairly soon afterwards, Pakistan responded with seven underground explosions. These actions prompted cynics in Britain to ask, if it is good for our security to maintain a nuclear deterrent, then why is it bad for India and Pakistan to take the nuclear option?

A fair question?

# Chapter 22

# The deployment of strategic nuclear weapons

A nuclear bomb requires a means of delivery in order to become an offensive weapon. In principle, it could be conveyed to the target in a vehicle, abandoned, then exploded by remote control. But there would be a high risk of interception, so it would be the method used only by terrorists.

## Skybolt — The weapon that never was

The Hiroshima bomb was discharged over its target after being conveyed there by a US B-52 bomber. After Britain had exploded its own bomb in 1952, a very similar one to the plutonium-based one used over Nagasaki, the plan was to accumulate a stock of bombs to be delivered to their targets by the V-bomber force then under development (see Chapter 6). In order to extend the life of this force, the stand-off bomb known as Blue Steel was developed. It was a liquid-fuelled rocket with a thermonuclear (H-bomb) warhead, as described in Chapter 7. As a stand-off bomb it would enable bomber aircraft to attack a target from some distance outside the enemy's main outer defences.

Taking this method of delivery to its ultimate, the US Air Force embarked on the development of the Skybolt rocket which would have enabled the planes to launch their missiles 1600 or more kilometres away from the target. The British government was attracted to Skybolt and an agreement was therefore drawn up with the US government to sell these missiles to Britain when available, the warhead of the weapon to be made in Britain and the government to have sole control over the use of the British missiles. Britain had been developing rockets for use as ballistic missiles to be fired from the ground but had cancelled Blue Streak, its leader in this field, in 1960, and so now all rocket propulsion work had ceased in this country. Britain's nuclear deterrent for the future was henceforth going to become dependent on getting Skybolt from the US. But President Eisenhower cancelled the Skybolt project in 1962 without informing the British government, much to its alarm and annoyance at not having been taken into the US government's confidence at the outset.

## The Nassau agreement

In December 1962, Prime Minister Harold Macmillan flew to meet Eisenhower's successor, President Kennedy, in Nassau, capital of the Bahamas, and it was there that he was made truly aware of British concern. But the President would no doubt have informed Macmillan that he had to contend with the views held by the US government and military lobby, as there was a feeling that Britain ought to phase out its independent nuclear deterrent. But there were still many members of the British government who wanted to keep it for prestige reasons and hence there was pressure on Macmillan to get an agreement at Nassau which would make this possible. He succeeded in this task.

The Nassau agreement meant that Kennedy would let Britain have Polaris, a surface-to-surface ballistic missile capable of being launched from submarines. In 1963, the Polaris sales agreement was signed between the two countries; it was the instrument that translated into working rules the broad principles agreed at Nassau. It has been said that Kennedy was working against his own foreign policy in signing this agreement since he wished to see Britain join the Common Market but now it would be likely that the Polaris deal would result in de Gaulle vetoing Britain's entry. In fact, convinced after the Polaris deal that Britain was subordinate to the US, de Gaulle did veto Britain's application to join the EEC.

When Labour came to power in 1964, Harold Wilson as Prime Minister continued the bi-partisan policy of the deterrent with the Conservatives and so Polaris stayed at the heart of Britain's nuclear weapons programme right to the Tory take-over of government in June 1970 and beyond.

## Origin of Polaris in the USA

Polaris was the name originally given by the US Navy to an intermediate range ballistic missile developed for launching from submerged submarines. Work on Polaris started in the US in 1955 and was completed in only five years, when the specially-constructed USS *George Washington* fired two shots from a submerged position off Cape Kennedy on 20 July 1960. Each vessel carried sixteen missiles in vertical tubes and the initial, A1 version used a two-stage solid-propellant rocket motor giving it a range of 2,200 kilometres. But the US Navy pressed ahead with improvements and these came in stages with the A2 following in 1962 and the A3 in 1964, each providing an increase in range, finally reaching to 4,600 kilometres. What the Americans were aiming to do was to develop a response to the anti ballistic missile(ABM) defences which the Soviets had installed round Moscow and the Polaris A3 was in strategic terms a multiple re-entry vehicle (MRV) equipped with three H-bomb warheads designed to knock out an ABM target.

This was known as 'counterforce' strategy, announced in a speech by Robert McNamara, US Defence Secretary, at Ann Arbor in 1962. Its further development continued during the 1960s when it became possible to add independently targetable warheads to the ballistic missiles; the weapons were then known as MIRVs. At the outset there was pressure from the arms industry in the US for the government to get involved in developing this technology and RAND, Aerospace Corporation, Lockheed and Lincoln Laboratories became the prime contractors for MIRV. Indeed, Lockheed had been pressing for a defence programme to replace Polaris A3 which was coming to the end of its development cycle in 1963.

## British nuclear submarines and Polaris

Basically there are two kinds of nuclear submarine — those which are nuclear powered, and those which are nuclear powered and *also* carry Polaris missiles.

The first nuclear-powered submarine in the world was the *Nautilus*, completed for the US Navy in 1955. Following the amendment of the US Atomic Energy Act in 1958 (see Chapter 12), a British team went to the US to learn about nuclear propulsion for submarines. This team was not popular in certain quarters; Admiral Rickover of the US Navy was lukewarm about the whole project but was overruled due to the pressure of the 'industrial complex' which wanted British contracts. Following the visit, Britain's first nuclear propulsion system was bought from Westinghouse and installed at Barrow in Furness by Vickers under the supervision of Rolls Royce and Associates, a company set up to buy the components from the US firm. Britain's first nuclear-powered submarine, the *Dreadnought* (3000 tons displacement), was completed in 1963 and went to sea with a crew trained in America. Several more of this type were built and the fuel, enriched uranium, came from the USA. The later ones had a speed under water of 30 knots, very valuable for what was essentially a 'hunter-killer' type of submarine. They were to be used to attack the USSR's submarines and protect British shipping. One of them sank the *General Belgrano* at the start of the Falklands War, using World War II torpedoes.

In parallel with the building of nuclear-powered submarines in Britain, a UK prototype nuclear submarine power reactor plant was built at Dounreay, incorporating British ideas and machinery. The plant went critical in the early 1960s and ran, intermittently, for twenty years, its main role being to train engineering staff in operating a reactor plant and also to test control systems. Four of the nuclear-powered submarines were specially built to carry the nuclear war-headed Polaris missiles and each one carried 16 A3s. Four Polaris submarines was the minimum number considered viable by the government's military advisers; it was to ensure there would always be at least one fully operational and in a secret position under the ocean all ready to fire off its weapons on a very short notice of command. The other submarines would either be travelling to and from base for fuel and stores replacement and crew change, or would be in dry-dock having a refit. In July 1969, the Royal Navy, with these four submarines, took over from the RAF the UK's contribution to the Western Strategic Nuclear Deterrent. The Royal Navy had been keen to get its hands on the British deterrent for some time and there had been tremendous inter-service rivalry in the 1960s.

Four Polaris-firing submarines were a very costly investment. The initial capital costs were probably about £850 million at 1963 prices, about 7% of the defence budget, and annual maintenance costs were estimated at £270 million at the very least, for other related costs were probably hidden in the defence budgets. William Crowe, a former chairman of the US Joint Chiefs of Staff, was writing a Ph D thesis in Britain in 1965 whilst he was still a Commander in the US Navy. In his thesis he expressed the view that the ballistic missiles on the Royal Navy's four Polaris submarines would be of marginal relevance. His view was that a country faced by severe economic problems such as Britain should be demanding more of a return on its investment of £300 million plus operating costs. 'The abandonment of nuclear weapons would release funds which could be profitably employed elsewhere,' he wrote. The entire Polaris force gave Britain an arsenal of 64 missiles, sufficient to destroy 30 cities in one 24-hour period, small though it was compared with that of the US Navy. But there was always a problem to be faced concerning their use; a ground-burst weapon or even one that went off very low by error could release enough radioactive particles to contaminate a large area of western Europe and it was even possible that the USA might have been affected.

## Chevaline

After the Conservatives came to power in 1970 they decided to give the front end of the Polaris missile advanced penetration aids and a more manoeuvrable payload. This was in anticipation of improvements in the Russian anti ballistic missile capacity — improvements which in the event did not materialise. This programme, known as Chevaline, was secretly approved by Edward Heath whilst Prime Minister in 1970. Labour returned to government in 1974 and continued with Chevaline but this was never announced. By 1981 the Polaris development costs had reached a billion pounds. For years it was disguised from parliament and the public, being hidden in the Defence Estimates, and even members of the Labour cabinet between 1974 and 1979 said later that they had been unaware of the project. In addition, it was secretly decided to spend at least £300 million to provide the Polaris missiles with new rocket motors.

But matters did not cease there, for it was announced in July 1980 (after the Tories had gained power under Margaret Thatcher) that the Polaris strategic force would be replaced by the more powerful Trident submarine-launched ballistic missile system, this again to be purchased from the USA starting in March 1982. At first it would use Trident mark I, but then the decision was made to go for the 'improved' version, Trident II (D5). This required a major modification of the submarine programme; four 14,480 tonne submarines were to be built in the UK, to be powered by British pressurised water reactor atomic propulsion engines and to have improved sonar systems. The capital cost of all this was believed to total £8,000 million by mid-1982.

One of the serious disadvantages with all nuclear submarines which used pressurised water reactor propulsion units was the decay heat, liberated for some time after the plant shut down. In an old core which has run for two years or more, this decay heat is considerable and it must be continuously dispersed if the core is not to melt down, which would create a really serious situation. So electric power must be available all the time in order to run the decay heat cooling pumps while the vessel is in harbour. In the 1990s after the break-up of the Soviet Union, several Russian nuclear submarines were laid up and became dependent on electric power supplies from an onshore station. When the Russian Navy could not pay their electricity accounts, the supply was therefore cut off and armed sailors had to compel the power company to restore power in order to avoid a melt-down catastrophe.

## Other British nuclear weapon systems

So far we have concentrated on strategic nuclear weapons, conveyed first by the V-bomber force and then by submarines using Polaris missiles. But other types of nuclear weapons were developed requiring the design of new types of nuclear warhead. At the height of the Cold War, NATO (in common with the Warsaw Pact) had several thousand nuclear weapons in Europe ready for all-out war. Directly under NATO command were Cruise, Pershing II and Lance missiles plus bombs, artillery shells and mines. Many were owned by the US but other countries in NATO, notably Britain, were allowed to operate them under a dual control system.

Britain also had its own so-called tactical nuclear weapons which were not integrated into the NATO command structure. These included nuclear depth bombs carried by Nimrod anti-submarine aircraft as well as artillery shells and Lance missiles but Britain did not have sole control over all of them; many were still under dual control with the US. Also, the RAF had over 100 British-made tactical nuclear warheads known as the WE177 by 1980, carried mainly by Jaguar and Buccaneer strike aircraft and, later, by the Tornado aircraft. However, for a sixteen-year period prior to 1978, Scimitars and Buccaneers operated in a nuclear-capable role on aircraft carriers such as *Eagle*, *Centaur*, *Victorious* and *Hermes*. Later, the Royal Navy had WE177s for its Sea Harriers operating from aircraft carriers and a depth bomb variant of the WE177 which could even be carried on helicopters deployed on many destroyers and frigates. The next chapter of this book deals with the history of the deployment of tactical nuclear weapons by Britain outside the European theatre, including their presence close to, if not actually in, a theatre of war.

## The production of nuclear warheads

The development and adoption of various alternative war-fighting strategies resulted in massive investment in nuclear arms which necessarily absorbed immense wealth in terms of both money and resources. Each increase was usually justified on the basis that the Soviets were in the lead and the Western alliance had to catch up and overtake them. But of course, in turn, the Soviets would then decide they just had to catch up with the Americans; that was what the arms race was all about. In fact, intelligence about Soviet military hardware was often wrong. Indeed, one school of thought believed that the American strategy was to push the arms race to the point at which the Russian economy suffered ruin. And the industrial military complex in the US plus the establishments concerned with nuclear weapons in Britain were only too ready to develop and produce more arms.

## The human risks

Although Britain imported conveyance systems from the US, especially rockets, the nuclear warheads were always developed and produced over here. The nuclear materials for these would travel by road or rail from Windscale (Sellafield) in Cumberland, across country to Aldermaston in Berkshire. As the opponents of nuclear weapon production have always been ready to stress, there was always the risk of an accident en route which could cause radioactive contamination in the surrounding area. At Aldermaston these materials would be fabricated into components for nuclear or thermonuclear warheads and then despatched the six miles down the road under heavy guard to the weapon assembly factory at Burghfield. This practice has continued for many years. Accidents (although of course no nuclear blasts) have occurred from time to time at these establishments and since radioactive materials were often involved, operating staff were liable to suffer radiation injury and, quite possibly, impairment of their health. In some cases, death followed in due course but its connection with the accident at work could always be very difficult to prove in court. However, over the years, a number of out-of-court settlements have been agreed for the payment of compensation to workers and families at Windscale and Aldermaston.

# Chapter 23

# Deployment of tactical nuclear weapons in crises

'The consciences of civilised nations must naturally recoil from the prospect of using nuclear weapons... We have to be prepared for the outbreak of localised conflicts short of global war. In such limited wars the possible use of nuclear weapons cannot be excluded.'

From the 1956 Defence White Paper.

Much of the previous chapter dealt with the history of *strategic* nuclear weapons in the context of the confrontation between NATO and Warsaw Pact forces, with the European theatre largely in mind. But brief mention was made of the development of *tactical* nuclear bombs: in this chapter we go on and examine British nuclear policy regarding their deployment in times of crisis, independent of NATO. There is little published material available in this area of study and the compilation of information has to depend mainly on interviews which have had to remain on a non-attributable basis.

Many people not professionally involved have tended to believe that nuclear weapons are simply the so-called 'ultimate' deterrent with no possibility of them being used in any situation short of World War III, which would in any case never involve the use of nuclear weapons because neither side would be mad enough to use them for fear of retaliation... So when the Cold War ended, many people became even more sure that there was no longer any question of nuclear weapons ever being used in a war situation. This view is unjustified because Britain has deployed nuclear weapons outside the NATO theatre since the middle 1960s, specifically in several conflicts involving non-nuclear powers. Indeed we have some indications of the presumed value and therefore possible use of nuclear weapon systems in a confrontation falling far short of a global war. Nuclear policy has gone beyond the notion of ultimate deterrence, moving on to actual planning and training for limited use by Britain in the belief that a nuclear war against a non-nuclear armed opponent could possibly be kept limited — and successful. We now discuss three theatres of war, all of which have occurred since the bombs were dropped on Japan in 1945 by the Americans with full British support.

## Confrontation in South-East Asia

Britain has deployed nuclear weapons as far afield from Europe as South-East Asia. In the mid 1960s, V-bombers were regularly based at Singapore, some of them nuclear armed. Even as early as the late 1950s there was a period of debate among defence chiefs about the possibility of waging a limited tactical nuclear war in the Far East. The Chiefs of Staff Committee (COSC) may well have thought that if a conventional war in the region resulted from China attacking Hong Kong or Formosa, or even one of the tiny off-shore islands, then nuclear weapons could be used without risk of the conflict escalating to the strategic level. Such discussions would surely have preceded knowledge of China achieving nuclear weapons. Later, discussions would have centred around the sending of nuclear forces to the Indian Ocean, providing a nuclear guarantee to India against China. The Wilson Government's Defence White Paper of 1965 was referred to by *The Observer* under the heading 'Labour's Bomb and the White Man's Burden.' Although it did not specifically say so, the white paper gave some indication that Britain intended to keep an independent nuclear force of V-bombers and carrier-based aircraft outside the NATO area; the main purpose from the government's point of view was quite probably to give a 'nuclear guarantee' to Commonwealth countries which feared China and possibly, in the future, Indonesia. Thus would Commonwealth countries be deterred from wanting to make their own nuclear bombs.

In early 1963, the possibility of nuclear deployment by Britain in Singapore did present an implicit threat to the Jakarta authorities during the Indonesian Confrontation. Air Chief Marshall Sir David Lee, historian of the Far East Air Force, has commented that

'the knowledge of RAF strength and competence created a wholesome respect among Indonesia's leaders and the deterrent effect of RAF air defence fighters, light bombers, and V-bombers on detachment from Bomber Command was absolute.'

In actuality, it might be argued that Canberra and Victor deployments to Singapore represented a conventional force deterrent but the very fact of their dual-capacity (high explosive or nuclear bomb capability) did allow Britain to imply a greater potential. A type of shadow blackmail?

## Nuclear weapons and the Falklands War

When Argentina invaded the Falkland Islands in April 1982 a naval task force was rapidly assembled, some elements setting forth by sea within four days. This was paralleled by a debate in parliament in which the defence secretary, John Nott, specifically stated that warships were

'sailing under wartime orders and with wartime stocks of weapons.'

When defence ministers were questioned on whether nuclear weapons were being deployed, they replied that there was no risk of escalation without actually saying that there were no weapons deployed — typical Whitehall phraseology! Writing in *The Observer* six days after the initial elements of the task force had departed, the paper's defence correspondent, Andrew Wilson, reported that nuclear weapons almost certainly embarked in some of its ships. Tactical nuclear naval weapons, atomic depth charges for use with Sea King helicopters and free-fall bombs carried by Harrier jump-jets were standard NATO equipment. Although Margaret Thatcher's war cabinet probably had no intention whatsoever of using these weapons in their war plan, some defence experts were concerned at what might happen if the conflict were to escalate in some unforeseen way. The very presence of nuclear weapons in a war zone involves serious risks even if they are not to be used.

At the same time as the task force was sailing out, a NATO exercise known as Spring Train was in progress in the Mediterranean. The destroyer HMS *Sheffield* and certain other ships were disengaged from this exercise and directly despatched to rendezvous with the task force at Ascension Island in the South Atlantic. Former navy minister Keith Speed MP, who had resigned during the previous year in protest at cuts in the navy budget, is on record as saying:

'I would have been astonished if those ships detached from Exercise Spring Train had not been carrying nuclear weapons.'

In fact, according to Tam Dalyell MP there was some consternation in the Ministry of Defence that such a large proportion of the Royal Navy's entire stock of nuclear weapons was heading for a probable war zone. The worry would have been twofold; not only concern over possible escalation to nuclear use but also the risk of losing such weapons in a conflict restricted to conventional weapons. Some MOD senior staff were reportedly very worried over losing any of the navy's relatively small stock of tactical nuclear weapons and Dalyell maintained that some of the latter, but not all, were lifted back by helicopter before the task force got further out than the Western Approaches.

According to a member of the Royal Fleet Auxiliary, some nuclear weapons were offloaded from task force ships onto the Royal Fleet Auxiliary *Regent* at Ascension Island. Although this ship continued on to the South Atlantic with tactical nuclear weapons on board it was kept well away from the main conflict zone although its sister ship RFA *Resource* actually was at San Carlos Bay at the time of the amphibious landings. The deployment pattern applied to nuclear weapons on the two carriers *Invincible* and *Hermes* is unknown.

The destroyer *Sheffield* may well have carried WE-177 nuclear depth bombs, although there is no proof of this, but there was speculation that the very extensive salvage operations conducted on the wreck of *Sheffield*, and indeed *Coventry*, were concerned with recovering these weapons. But the latter ship sank in only 30 minutes after being bombed and as there may well not have been time to destroy some advanced equipment including code systems, the recovery may have concentrated on retrieving the latter; *Coventry* may not have carried any nuclear weapons. The case of *Sheffield* may well have been different, it may have been carrying nuclear depth bombs, but it may be many years before the true facts are released.

To sum up, although elements within the navy may have been prepared to see *tactical* weapons deployed directly in that regional war zone, wiser counsel prevailed and indeed some nuclear weapons were kept at a distance. The reality was, it was a relatively short war and despite some unexpected British losses, it was unlikely that the MOD would have got around to seriously considering firing a tactical weapon before it all came to an end.

But what of the British *strategic* nuclear weapon — Polaris? A number of sources have maintained that a Polaris submarine was sent to the South Atlantic as far as Ascension Island. This major nuclear threat might just conceivably have been used if any of the task force's capital ships such as the *Canberra* had been lost in a missile attack. Used against what target, though? Cordoba in northern Argentina has been advanced as a possibility. If the Argentine air force had had rather less bad luck with its Exocet missiles and military defeat had been staring Britain in the face, would the Polaris submarine (if it was in position) have been ordered to fire in order to save the Thatcher government from collapse? Indeed this question has been seriously raised as a possible scenario but the present writer cannot accept that saner and more humane counsels would not have prevailed in such a situation. We may well have to wait for thirty years to find out just what stratagems were considered by the British war cabinet during the Falklands crisis.

## The Gulf Crisis and War of 1990-91

After the Iraqi invasion of Kuwait in August 1990 and in the run-up to the war commencing in January 1991 there was much concern that Iraq might use chemical weapons (CWs) against the coalition forces. A number of western political sources hinted that any use of CWs by Iraq would get a nuclear response. Junior ministers spoke of 'massive retaliation' against Iraq if CWs were used and this surely could only have meant a nuclear response since Britain did not have CWs. In a report in *The Observer* on 30 September, a senior army officer of the 7th Armoured Brigade, which was on its way to the Gulf, confirmed that any Iraqi chemical attack on British forces would be met with a tactical nuclear response. Even the Minister of Defence, Tom King, said on television in November that the use of chemical weapons by Saddam Hussein would be 'the stupidest thing that he could do,' although he refused to be specific about just what Britain would do.

The US had tactical nuclear weapons capability in the form of aircraft carriers in the immediate vicinity of Iraq but the deployment of nuclear weapons to the region by Britain presented difficulties. Although there were strike aircraft at Bahrein and Oman, basing nuclear weapons there would not have been acceptable to those states and the same would have applied to RAF Akrotiri in Cyprus. The weapons would have to be shipboard. This would not have been easy, in terms of quick response. Prior to the actual war, it is probable that various plans were devised, such as converting the RFA *Argus*, a 28,000-tonne auxiliary vessel officially classified as an aviation-support vessel, to carry tactical weapons. Although *Argus* went to the Gulf during the crisis, there is no positive indication that it carried nuclear weapons. Another possibility which would have been contemplated would have involved moving nuclear bombs for Tornado strike aircraft from Britain or Germany. If the Iraqi forces had used chemical weapons on any large scale, the most likely response would probably have involved the use of free-fall nuclear weapons from American F-15s or F-18s with political and practical support from the British government; shades of August 1945? But this is purely academic, in the event no large-scale use was made of CWs by Iraq.

## In conclusion

In this chapter we have described three theatres of war directly involving the UK since World War II, completely outside the potential area of conflict between NATO and Warsaw Pact forces, where nuclear weapons were at least considered for use by the UK if not actually deployed. Nothing has ever been officially admitted about this by British governments, such is the power of our security system with its Official Secrets Acts, D-notices etc. But with the ending of the Cold War, the utility of nuclear weapon use against Third World countries must in all probability receive consideration in UK defence circles. As we shall see in Chapter 24, the Trident force, which followed Polaris, is to be maintained in service as a potential strategic weapon even though the Cold War is over. Also, some of the Trident missiles may be adapted for use as tactical nuclear weapons when all the WE-177 nuclear bombs have been removed from service.

*Footnote*

The author is indebted to Professor Paul Rogers of the University of Bradford for kindly supplying some of the information used in this chapter.

# Chapter 24

# The Cold War ends: What price strategic weapons?

When in late 1979 the energy secretary was preparing his proposals for nuclear power, his cabinet colleague Francis Pym, defence secretary, would have been studying the latest nuclear war-fighting policy inherited from James Callaghan's Labour cabinet. This was apparently going to include playing host to US Cruise missiles. These are small jet-propelled missiles flying at about the speed of a jumbo jet and capable of travelling over 1,250 miles. They are fitted with a special computer to guide them, using a map of the ground to be covered. Each missile, if fitted with a nuclear warhead, could create an explosion about fifteen times greater than the Hiroshima bomb. At times of tension, they would have been sent from their bases and spread around the country. The only way to be certain of destroying them would have been a heavy nuclear counter-attack killing millions of people and making large tracts of land uninhabitable. They were thought to be necessary as a counterforce to the Soviets' SS missiles.

Although it would be well-nigh impossible to prove, the US may well have exerted leverage over the British government in this matter; to place the country under such a terrible risk could not have been a decision easy to make. It was in June 1980 that Pym finally named Greenham Common and Molesworth as the sites for Cruise.

In the following month, he formally announced the replacement of Polaris by the more powerful Trident submarine-launched ballistic missile system. The Polaris force of four submarines, each carrying sixteen missiles, was to be replaced by Trident in the mid 1980s. Trident would have a range of 6,000 miles, more than double that of Polaris, and would be harder to detect, thus cutting down on warning time. Trident was more than a replacement for Polaris, it would increase Britain's strategic nuclear force tenfold or more, thus giving the Russians an excuse to increase their nuclear forces to catch up. The *initial* cost of Trident was estimated to be £10 billion but in the event this will have increased considerably as also will its running costs.

There was never any argument in parliament over these two decisions — Cruise and Trident — owing to the continuing cross-bench agreement on nuclear weapons between Labour and the Tories. However, many people in Britain were becoming worried about the coming of Cruise missiles. In October 1980, the revival of CND was marked by a march of 50,000 people to Trafalgar Square. In the following October there were said to be 150,000 anti-nuclear demonstrators in London; similar large protests were taking place throughout European cities where Cruise was to be based. Women's groups in Britain set up permanent protest camps near the gates of Greenham. Camps and frequent demonstrations also took place at Molesworth. Undeterred, on 13 November 1983 Michael Heseltine announced the arrival of Cruise missiles at Greenham Common.

The early eighties were said to be the height of the Cold War. Public demonstrations in Britain against nuclear arms were frequent, not only at nuclear weapon stations and the Faslane nuclear submarine base, the latter also having a permanent peace camp, but also at the early-warning stations of Fylingdales and Menwith Hill, both situated in North Yorkshire. People from religious and environmental bodies as well as political groups penetrated these high-security sites or chained themselves to the perimeter fences so that they could be brought before magistrates courts, there to make their protests known.

Expenditure by governments East and West on preparations for war reached new heights. Nevertheless by 1986 there was a growing feeling in the air that *rapprochement* between NATO and the Warsaw Pact forces was not only feasible but imminent.

## The end of the Cold War

In October 1986, in an atmosphere of global suspicion, Ronald Reagan and Mikhail Gorbachev surprised everyone by meeting together in Reykjavik in a cordial atmosphere. This date has been regarded by some as the beginning of the end of the Cold War. It may well have been, but there was a long way to go. The peace movement saw the opportunity to push for progress by way of arms control negotiations in Europe as well as unilateral initiatives, for public opinion was now very much in favour of advancing rapport between West and East. It is interesting that the Cold War began to decline at a time when hawks like Reagan and Thatcher were leaders in the US and UK respectively. So was it largely because of Gorbachev's realisation that his side could not go on indefinitely investing in yet more arms because of the harmful effects it was having on the Soviet Union's economy in general? Whatever the answer to that question might be, Gorbachev now felt that the time had come to show positive movement towards peace with the NATO powers and he was personally responsible for unilaterally withdrawing 500,000 servicemen, 8,000 tanks and 800 aircraft from the Warsaw Pact front lines in Europe.

The next salient step in the peace process came with agreement over limitations to *intermediate range nuclear forces*. The so-called INF Treaty was signed in Washington DC on 8 December 1987 in an atmosphere of celebration and optimism. Although the weapons to be scrapped accounted for only about 3 per cent of the world's nuclear arsenals, it was a serious beginning. Significant numbers of modern so-called counterforce nuclear missiles were involved. The agreed procedures for verification and for destroying the weapons and their support systems created important precedents for the future and prospects for growing East-West detente were improved.

Although in the early 1980s the NATO leaders had argued strongly that Cruise and Pershing missiles were essential for West European security, following the INF Treaty they now accepted that these missiles were expendable. However, NATO leaders were slow to relinquish their belief in the need for battlefield nuclear weapons. But the idea of the creation of a Nuclear Weapon Free Zone (NWFZ) across Europe was slowly gaining support and this, if implemented, would result in the removal of many battlefield nuclear weapons. For reasons of verification, the agreement would need to include dual-capable systems including the Warsaw Pact FROG and NATO's Lance. If and when the zone was created it would clearly open the way to further arms control and disarmament measures. But such changes would take a long time to get onto a firm agenda and then be successfully negotiated.

However, any case for keeping Cruise missiles on British bases had now evaporated and it was only a matter of time before they were removed back to America.

## The London Declaration, July 1990

NATO, including the UK, were certainly 'adjusting' their nuclear doctrines and postures by the end of the 1980s. Indeed NATO changed its policy to reflect 'a reduced reliance on nuclear weapons'. Then in the so-called London Declaration emerging from the NATO summit conference held in London, 5-6 July 1990, it was declared that nuclear weapons were now to become 'weapons of last resort'. Thus NATO was tacitly admitting, publicly, that previously they would have been used as part of a deliberate war-fighting strategy in the event of open conflict between NATO and Warsaw Pact countries.

The main practical change in the UK's nuclear posture at about this time was the decision to begin to phase out tactical nuclear weapons. In Chapter 23, we referred to the probability that tactical nuclear weapons were considered for deployment in the Gulf War of 1990-91, although this was thought to be difficult in practice. Henceforth naval tactical nuclear weapons began to be withdrawn from service. By 1994, Britain had halved its arsenal of air force tactical weapons and planned to have the remainder withdrawn — and dismantled — within a few years, thus leaving it deploying only one nuclear weapon system (Trident) for the first time since the 1950s. The government had hitherto considered procuring an air-launched stand-off nuclear missile to replace the tactical nuclear weapons being withdrawn but this development — known as TASM — was cancelled in 1993. Instead, the government decided to reduce the number of warheads on some of the Trident missiles, these thus becoming few or single-warheaded *sub-strategic missiles*.

So far as the *strategic Trident missiles* were concerned, the government announced in its Defence Estimates for 1994 that each submarine would carry no more than 128 warheads and on the basis of the government's 'present assessment of its minimum deterrent needs', the limit may in fact only be 96. But it still remained government policy not to reveal precisely how many warheads would be deployed within these limits. Finally, the British Prime Minister, John Major, informed President Boris Yeltsin that British strategic nuclear weapons were no longer targeted on Russia.

## The Nuclear Non-Proliferation Treaty and the Comprehensive Test Ban Treaty

Following the deliberations in NATO on nuclear posturing after the end of the Cold War, and Britain's own decision-making about nuclear weapon development and deployment for the future, two important developments took place in the international sphere.

In Chapter 21 we described the setting up of an international non-proliferation treaty which came into force in 1970. The number of nations signing this treaty eventually reached 167. But it had to be reviewed after 25 years and so the '1995 Conference of the Parties to the Treaty on the Non-Proliferation of Nuclear Weapons' duly took place at the United Nations in New York from April 17 to May 13 of that year. The Conference agreed to:

• extend the Non-Proliferation Treaty (NPT) indefinitely;
• adopt a set of principles and objectives on non-proliferation; and
• create an enhanced review process.

However, on May 15, just two days after the conference ended, China conducted a nuclear test explosion and France announced plans to carry out a series of eight tests between September 1995 and May 1996. For a time it seemed that the US was also planning to resume testing but that threat receded. Governments, religious bodies, groups and individuals around the world criticised the Chinese and French governments. Indeed, French goods and services were boycotted in many countries. But the UK government avoided criticising the Chinese and French.

This was the first NPT conference in which all five nuclear weapon states participated — China, France, Russia, the UK and the USA. It was inevitable, given the diversity of interests of all the countries participating, that progress would be slow and unsure. It was mainly due to the initiative of South Africa, the adroit chairmanship of Ambassador Dhanapal of Sri Lanka and good work behind the scenes by Mexico and a few other countries that the conference finally agreed the three main principles listed above. Also addressed were the universality of the treaty, nuclear weapon free zones, security assurances and safeguards. While up to that time a review conference had been held every five years, it was now agreed that preparatory committees would meet in each of the three years preceding them. This would result in the nuclear weapon states being under more pressure to show progress towards real nuclear disarmament under Article VI.

At the NPT in 1995 it was agreed that states would negotiate and complete a Comprehensive Test Ban Treaty (CTBT) by the end of 1996. In order to reach this deadline a final text had to be in place by June 1996, to be submitted to the conference on disarmament in Geneva and finally the UN General Assembly in September. This was a tight schedule to work to but the UK, US and France did make some helpful technical concessions before the end of 1995. In particular they accepted that the concept of 'zero testing' would even include a ban on minor test explosions. At the end of January, France announced an end to its nuclear testing programme. All then went well enough for the world's five declared nuclear weapon powers to be able to sign agreement to the prepared draft for the CTBT by the summer. President Clinton was the only head of state there to sign but Malcolm Rifkind, the UK foreign secretary, and his counterparts from France, China and Russia very quickly followed Mr Clinton's action. But it required 44 states to ratify the treaty before it would take effect and full verification could begin.

The treaty banned all nuclear test explosions, of which there had been over 2,000 since the first at Los Alamos, New Mexico, in July 1945. Some analysts complained that nuclear countries only agreed to the treaty because of having sophisticated computer simulation techniques available. Other experts said it was a grave exaggeration to say that this could replace actual testing, for computers were not that good! Anti-nuclear groups hailed the CTBT as a landmark in the history of arms control and environmental protection and said that it could actually hasten moves towards disarmament by the big powers. Greenpeace called the signing an 'historic moment', coinciding with the organisation's 25th anniversary; it was originally founded in protest at nuclear testing by the US off Alaska. Thomas Graham, the US disarmament envoy, said:

> 'The moral force of signature will create a norm of international behaviour which will ensure that the Chinese test in July 1996 will be the last nuclear test carried out anywhere by anybody.'

Well, we shall see...

## A final note on British nuclear arms policy

In Chapter 7, I referred to the moment of hope in 1955 when the world seemed to be so near to realising the beginning of general disarmament negotiations and the international control of nuclear weapons. It would be nice to be able to conclude by saying that after all the perils endured and the horrendous waste of resources and damage to the environment suffered during the succeeding 40 years, the world was now about to pick up the thread again and move on to make enduring progress towards the elimination of nuclear weapons and accompanying this, the achievement of real peace. Of course we cannot know this, but the author hopes to see real progress made during the few remaining years of his life.

The book is largely devoted to nuclear politics in Britain. Having signed the two treaties described above, Britain, as a declared nuclear state, was committed to giving a lead in achieving world nuclear disarmament, especially by example. What chance was there of our country playing its part? Consider three quotes from published statements by political parties:

> 'National nuclear capabilities continue to underpin British defence strategy and provide the ultimate guarantee of our security.' (From the Conservative government's 1995 Statement on Defence Estimates.)

> 'Retain UK nuclear weapons as long as other countries possess them... Restrict these to 3 or 4 Trident submarines with no increase in warheads over the Polaris figure of 192.' (From a Labour Party policy document dated February 1992.)

'The UK should retain an independent nuclear deterrent ... as an insurance against unknown risks... This can be provided by the Trident weapon system.' (From Liberal Democrats Policy Paper No 6, September 1994.)

From these and many other statements it was clear that the cross-party agreement on nuclear policy was likely to continue, as it has done for the last half century or so and even since the 1997 general election. But there are in the Labour and Liberal Democrat parties small but vociferous groups opposed to Britain continuing to hold nuclear weapons. The diminutive Green Party is united in its opposition to nuclear weapons — and nuclear power generation; but it is unlikely to have a voice in the British parliament in the absence of electoral reform.

So far as UK public opinion in general goes on nuclear weapons, it does not seem to have changed much since the Cold War ended. A 1995 opinion poll found 23 per cent wanting to keep nuclear weapons for the foreseeable future with 59 per cent believing the UK should dismantle nuclear weapons gradually in coordination with the other nuclear states. So it rather looks as though the urge for ridding the world from nuclear armaments will need to come from outside. Indeed it may well come from an unexpected quarter.

On 5 December 1996, 60 generals and admirals from many different countries, meeting in London, called for the elimination of nuclear weapons as they represented a clear and present danger to the very existence of humanity. They shared with Field Marshal Lord Carver, one-time chief of Britain's defence staff, the view that a nuclear deterrent is riskier than not having one. The gathering also included two former NATO supreme commanders, John Galvin and Bernard Rogers, as well as Russia's General Alexander Lebed, Yeltsin's former security adviser. Their statement proposed three moves: further large cuts in nuclear stockpiles, taking those that remain gradually off alert, and declaring that the world must now work towards their total elimination. One obvious prerequisite for the latter is a worldwide system of inspection to ensure that rogue states or terrorists cannot acquire such weapons. With this would go 'an agreed procedure for forcible international intervention.' The generals concluded:

'The end of the Cold War makes all this possible. The dangers of proliferation, terrorism and a new nuclear arms race render it essential.'

If statements like this can be made by people like that, then it would seem to the present writer that it is high time for following their lead. But it would seem that Britain is not doing so.

In July 1998, the Blair government's Strategic Defence Review stated that Britain would continue its strategic nuclear deterrent based on the four Trident submarines, one of which would always be on permanent patrol. Its number of warheads would be 48, representing a reduction of 70 per cent in explosive power since the end of the Cold War. But explosive power is not the same as destructive power; compared with Polaris, the Trident weapon is about four times more accurate, has twice the range and possesses independently targetable warheads. However, the government has announced that the WE177 free-fall nuclear bombs have been withdrawn from service.

Britain's present intention to continue with nuclear weapons production entails the continued procurement of materials for warhead manufacture. This topic is followed up in the next and final chapter of this book.

# Part 4

# THE NUCLEAR
# LEGACY

# Chapter 25

# The long-term costs of the nuclear age

'We must assume that these wastes will remain dangerous and will need to be isolated from the biosphere for hundreds of thousands of years. In considering arrangements for dealing safely with such wastes man is faced with time scales that transcend his experience.'

From the report of the Royal Commission on Environmental Pollution, September 1976, *Nuclear Power and the Environment*, Cmnd 6618 ('The Flowers Report')

In Chapter 4, we observed that although it was easy enough for Clement Attlee and one or two of his cabinet ministers, under the cover of secrecy, to go headstrong into the nuclear business, it would prove much more difficult for his successors to get out of it. Furthermore, Attlee and Co. could have had no idea that their successors would have to deal with what we might call the nuclear legacy at the close of the nuclear project. The decline of nuclear electricity generation, the end of the Cold War leading to a fall in demand for nuclear warheads and the concomitant need to decommission nuclear weapons are all bringing the nuclear powers face to face with costly and lengthy problems of health and safety in dealing with radioactive waste materials. In this final section of the book we take a brief look at this legacy, which present and future generations have to deal with.

## Dealing with high level waste

Dealing with the highly radioactive residues from chemical reprocessing is undoubtedly the most challenging problem of the legacy. In Chapter 17 the problems of storing high level waste (HLW) in Building 215 at Windscale were described. In the late 1950s, the AERE Harwell had recognised the need to reduce the bulk of HLW for storage by conversion to the solid state and in a form that could be safely cooled and stored in a suitable place. However, the work was stopped in 1960 because UKAEA headquarters said enough was known about dealing with HLW. Indeed, fifteen years later, the Flowers Report said that neither the UKAEA or BNFL had given the Royal Commission any indication that they regarded the search for a means of dealing with HLW as at all pressing.

The government had to respond to the Flowers Report. Its reply promised positive action in a paper, Cmnd 6820, published in May 1977. One result of this was the appointment of a standing Radioactive Waste Management Advisory Committee (RWMAC) in May 1978. But before then, BNFL had got the message and was already working hard on the HLW disposal problem and had also enrolled the assistance of Harwell — but many valuable years had been lost.

Other countries were also tackling the problem but it was not proving easy to find a suitable material in which to safely enclose HLW in a solid state. The Australian Defence Research Centre reported in 1980 that the intense radioactivity rendered boro-silicate glass highly susceptible to chemical attack and breakdown by moisture. In November 1980, BNFL announced that it intended to adopt the French AVM process for solidifying HLW wastes because it was considered more flexible than the British process. But in December 1983, the UK and Australia signed an agreement on collaboration in research into methods of solidification of HLW, concentrating in particular on the Australian 'Synroc' method. Back in June 1982, BNFL had announced that it was ready to seek planning approval for the building at Sellafield of a vitrification plant — a process for conversion of HLW to solids which could be safely stored for many hundreds of years without any risk of radioactive elements leaking out.

## Storing treated high level waste

Even when it becomes possible to turn the HLW into a satisfactory solid state, it still does not solve the problem of how, and where, to safely store this without any radioactivity leaking into the environment. In December 1981, the British government made it clear that it favoured the method of eventual deep burial of solid HLW but not until above-ground storage in solid form had been carried out for perhaps 50 years. In 1982, the government issued a White Paper, *Radioactive Waste Management* (Cmnd 8607), which announced the setting up of the Nuclear Industry Radioactive Waste Executive (NIREX). NIREX was primarily faced with the difficult problems associated with storage; a number of sites were investigated from a geological aspect but always there were other problems, notably the resistance of the local community, council or industry. Eventually it was realised that no one could be sure of the result of deep storage and so NIREX proposed that a laboratory should be established at a considerable depth to study the geology, temperature and water flow features at such a depth. It would be known as a 'rock laboratory' and in the summer of 1994, NIREX duly applied for permission to excavate for one at Sellafield, in the rocks deep below the reprocessing plants. In August 1994, a consultation paper, Radioactive Waste Management Policy, was issued by the Department of the Environment. It admitted that cost and safety problems were going to delay the building of an underground nuclear waste repository by up to 50 years, although the government still favoured deep burial as a long-term solution but 'no fixed deadline should be set for the completion of the process.' NIREX concurred with this view. All this was clearly going to cost a lot of money and a total figure of £7.5 billion was the estimate given at that time. But early in 1997, John Major and his cabinet decided that the rock laboratory at Sellafield must not go ahead — NIREX was back to the drawing board.

This is where the history of radioactive waste disposal comes to an end, for all else is pure speculation. Nevertheless it is tempting to leave the subject by posing two questions about the future of nuclear repositories. How can you warn people, many generations hence, that a nuclear waste repository was created in their neighbourhood many years previously? For it is quite likely that about 300 years hence, all records about it are liable to have been lost. And how will the local inhabitants know that their water supply may well be radioactive, be it from spring, well, or deep aquifer? We cannot begin to speculate about the type of human civilisation there may be on earth at such a distance in time, or even whether there will be any form of human existence at all. This is to say nothing about all other living things and our responsibilities to them.

## The legacy of obsolete products from the nuclear age

We have implied that there will be a legacy of derelict nuclear power stations and an accumulation of obsolete nuclear warheads, all to be dealt with safely. Decommissioning of nuclear plants is an art and a technology about which so little is known, although in Britain, useful knowledge is now being gained from the dismantling and stripping down of the Windscale AGR reactor (Chapter 13). But information is unlikely to be made available about the decommissioning of nuclear missiles and their warheads for very obvious reasons. In each case, after decommissioning, all radioactive and fissile materials need to be separated from the non-nuclear hardware. Unfortunately, lack of knowledge of the decommissioning operations and separation stages reduces estimates of costs down to pure guesswork. Some early estimates were quoted in the previous chapter but these are likely to be greatly exceeded. The government will have to pay in the end, but it will be a long drawn out business. Indeed, one of the problems is an ethical one; how should the costs be shared as between generations? The economists are unable to tackle the problem of intergenerational costs.

In the end the problems will remain of dealing with the separated waste fissile materials and radioactive isotopes. One would hope that the use of chemical reprocessing would be avoided for there is already a huge backlog of radioactive liquid wastes stemming from the reprocessing of spent nuclear fuel elements, said to be over 50,000 cubic metres, and this will grow so long as the reprocessing plants at Sellafield, Dounreay and Chapelcross continue to operate. Dealing with fissile materials present in unused or partly-used fuel elements, or nuclear warheads, poses serious problems. The accumulation of plutonium, sometimes referred to as the plutonium economy, will add to the risks of clandestine nuclear weapon production and trade. The idea of building nuclear reactors in which to use up extracted fissile materials has been quite seriously put forward. Indeed, BNFL has already completed a plant at Sellafield which can mix plutonium with uranium to produce a nuclear fuel which may be burnt in certain types of existing reactor. But, as we have already seen, there is no final solution in sight as yet to the problem of how to deal with spent nuclear fuel.

There is no money to be made out of all this, for there is no marketable product, if we exclude an international black market for the growing plutonium and U-235 stocks. These stocks are supposed to be monitored by the International Atomic Energy Agency (IAEA) but there are opportunities for stealing small stocks of weapons-grade material at various points of the nuclear cycle. Even so-called non-weapons-grade plutonium can be made into a nuclear bomb.

## Final comment

In Chapter 2 we described how the MAUD Committee was set up by the government in 1940 to report on the feasibility of Britain harnessing the energy of nuclear fission in the form of a bomb. The MAUD Report to the Ministry of Aircraft Production in July stated, *inter alia*, the following:

'There must always be a very close relationship between the exploitation of nuclear energy for military explosive purposes and for power production in peace and war. Development of one will have a considerable effect on the development of the other.'

I hope that my book will have convinced the reader of the truth of this very important proposition, for it means that if we are ever to abolish nuclear weapons, we have to be prepared to forgo nuclear power as a source of electricity, on our present knowledge.

Biotechnology now stands in a similar position to that of nuclear technology nearly sixty years ago. The genetic engineering that promises to do so much to improve the health and well-being of people everywhere is the same science and technology that can be exploited to develop biochemical arsenals of mass destruction. In tackling this awesome dilemma can we not benefit from our knowledge of what happened to nuclear technology.

# APPENDIX I

# Glossary of some key technical terms

Use of upper case letters signifies an entry in the glossary. A symbolic diagram of a nuclear reactor appears on page 70.

ADVANCED GAS COOLED REACTOR (AGR)   A carbon dioxide gas-cooled nuclear reactor, graphite moderated, using slightly enriched uranium oxide fuel contained in stainless steel cans.

ALPHA PARTICLE (α-particle)   The NUCLEUS of the helium atom consists of two PROTONS and two NEUTRONS in combination and thus has a double positive electrical charge.

ALPHA RAYS (α-rays)   Streams of ALPHA PARTICLES.

ANTI-BALLISTIC MISSILE (ABM)   The ABMs were weapons installed by the Soviets during the Cold War to provide a defence round Moscow against any possible attack by BALLISTIC MISSILES.

ATOMIC WEAPON   see NUCLEAR WEAPON.

BALLISTIC MISSILE   A warhead usually projected out of the earth's atmosphere by a rocket, re-entering near to the target, to which it is guided by remote control.

BETA RAYS (β-rays)   Beta rays, otherwise known as ELECTRONS, constitute a form of RADIATION.

BOILING WATER REACTOR (BWR)   A nuclear reactor which uses ordinary water as both COOLANT and MODERATOR. The water boils, under pressure, producing steam for driving a turbine. The uranium fuel is ENRICHED.

CHAIN REACTION   This is when an atomic NUCLEUS fissions following capture of a NEUTRON resulting in the release of more neutrons which in turn produce fission in other nuclei and a self-propagating process results, all without any external input of energy.

CHEVALINE   A secret British programme, commenced in 1970, for giving the front end of the Polaris missile more penetrative capacity and manoeuvrability.

COMMISSIONING   Bringing a new reactor system into operation. The final stage involves certain acceptance tests, one of which might mean running trouble-free for a specified number of hours.

COOLANT   A fluid which circulates continuously through the CORE of an operational reactor to extract heat.

COOLING POND   A large tank of water where SPENT FUEL elements are kept immersed to allow the decay of some of the RADIOACTIVITY resulting from fission in the reactor and to keep them from overheating.

CORE    The heart of a nuclear reactor where the fuel elements and usually a MODERATOR are located

CRITICAL MASS   The smallest quantity of FISSILE material in which a CHAIN REACTION can be sustained.

CRUISE MISSILE    A missile which travels in the earth's atmosphere like an unmanned aircraft but which has a predetermined flight at the end of which its warhead explodes. The latter may consist of a conventional high explosive or a nuclear one.

DECAY HEAT   Heat given out due to RADIOACTIVITY; see COOLING POND.

DECOMMISSIONING   Dismantling and safely disposing of a nuclear reactor or other RADIOACTIVE equipment at the end of its useful life. Problems in dealing with radioactive areas make it a very slow and costly process.

DERATING   Setting a power station to operate at a lower power output than originally intended.

DEUTERIUM   An ISOTOPE of hydrogen, sometimes called 'heavy hydrogen'. (See HEAVY WATER.) Has a PROTON and a NEUTRON in its NUCLEUS.

DESIGN RATING   The power output at which a nuclear reactor, or generating station, has been designed to operate.

DIFFUSION PLANT   A chemical processing plant used to increase the proportion of U-235 isotope in NATURAL URANIUM by passing the latter in a gaseous state, uranium hexafluoride, through a series of fine membranes. The process is known as ENRICHMENT.

DOUBLING TIME   A doubling time is the number of years it would theoretically take one FAST BREEDER REACTOR to breed enough plutonium to provide the CORE of a second one whilst still keeping itself fuelled.

EFFICIENCY    In reactor technology, it is the ratio between the useful power extracted and the potential energy release by the reactor.

ELECTRON   One of the fundamental particles of the atom, it has a negative electrical charge and a very low mass. See BETA RAYS.

ENRICHMENT    Increasing the proportion of fissile material in URANIUM (U-235) above its naturally occurring level of 0.7 per cent.

FAST REACTOR (FAST BREEDER REACTOR)  The NEUTRONS produced by fission are not slowed down by a MODERATOR, which is replaced by a surrounding blanket of FERTILE MATERIAL used for breeding PLUTONIUM. (But see DOUBLING TIME.)

FERTILE MATERIAL   Uranium depleted in the fissile isotope, U-235. (See FAST REACTOR.)

FISSILE MATERIAL, FISSILE ISOTOPE    Material capable of undergoing FISSION by NEUTRON bombardment, e.g. the isotopes uranium-235 or plutonium-239.

FISSION   The splitting of an atom's NUCLEUS occurring either spontaneously or by capturing a particle such as a NEUTRON.

FISSION PRODUCTS    Products resulting from the splitting of a heavy element through FISSION, including those resulting from the subsequent RADIOACTIVE decay of the initial fragments.

FUEL CYCLE   The series of operations through which nuclear fuel passes, from the mining of URANIUM ore to the final disposal of RADIOACTIVE wastes.

FUEL ELEMENT   Individual section of FISSILE MATERIAL in the CORE of a NUCLEAR REACTOR, enclosed in a metal alloy can.

FUEL ROD   Early name given to FUEL ELEMENT.

FUSION    The enormous force of the H-bomb is due to the fusion of atomic NUCLEI rather than their fission.

GAMMA RAYS ($\gamma$-rays)   Electromagnetic waves, similar to the well-known X-rays, but shorter in wavelength and very penetrating in character.

HALF LIFE    The time taken for a radioactive ISOTOPE to decay to half its original level of activity.

HALF LIFE PERIOD   (See HALF LIFE.)

HARVEST PROCESS   One of the processes developed in an attempt to store HIGH LEVEL WASTE in the form of an insoluble glassy material. (See VITRIFICATION.)

HEAD END PLANT   A plant for turning SPENT OXIDE FUEL into a product which could then be fed into a conventional REPROCESSING plant.

HEAVY WATER   Like ordinary water in most respects but the normal hydrogen atoms have been replaced by a heavier ISOTOPE called DEUTERIUM, which has an additional NEUTRON in its NUCLEUS. HEAVY WATER is a good MODERATOR but is very costly owing to the high consumption of electricity used in its production.

HIGH LEVEL WASTE (HLW)   Dangerous radioactive material separated off in the early stages of the chemical reprocessing of SPENT FUEL and which requires constant cooling owing to the excessive heat generated by radioactive disintegrations.

HIGH TEMPERATURE REACTOR (HTR)   A thermal reactor designed to work at CORE temperatures of about 1000deg C in order to greatly raise the temperature of the steam going to the turbines and thus increase the efficiency of electricity generation. It uses a graphite MODERATOR, a highly refractory fuel and an inert COOLANT gas, usually helium. 'DRAGON' was a successful joint OEEC prototype built in Dorset.

INTERMEDIATE LEVEL WASTE (ILW) Dangerous radioactive waste which does not require artificial cooling to prevent overheating.

IONISING RADIATION A general term including electromagnetic rays such as X-rays, GAMMA RAYS, ALPHA and BETA PARTICLES, all of which produce electrically-charged particles as they pass through matter. May be harmful to human tissue depending upon power level and contact time.

IRRADIATED FUEL — see SPENT FUEL

ISOTOPES Atoms which are chemically identical but have slightly different atomic masses due to variation in the composition of their nuclei. May or may not be radioactive.

LIGHT WATER Refers to ordinary water of high purity and low mineral content when used as MODERATOR in certain reactors.

LOAD FACTOR The amount of electricity actually generated by a power plant compared with its full load capacity.

LOW LEVEL WASTE (LLW) Assorted solid waste material containing low levels of radioactivity which could not be legally placed in any waste bin.

MAGNOX A special alloy of magnesium developed as cladding material for the URANIUM metal fuel used in the first CEGB nuclear electricity programme. Gave its name to the species of reactor stations.

MAGNOX REACTOR — see MAGNOX

MODERATOR Material inserted into the CORE of a THERMAL REACTOR to reduce through collisions with its atomic nuclei the velocity or energy of the NEUTRONS ejected by fission. The slowed-down NEUTRONS are more likely than the fast neutrons to undergo capture and thus promote further fissions.

MRV, MIRV Land missiles with multiple nuclear warheads, fired to leave the earth's atmosphere and re-enter it near its target. Specifically designed to attack a defensive system around Moscow which consisted of ANTI-BALLISTIC MISSILES (ABMs.)

NATURAL URANIUM Uranium as it is extracted from URANIUM ORE. Mainly consists of the FERTILE isotope U-238 but also contains 0.7 per cent of the FISSILE isotope U-235.

NEUTRON One of the fundamental particles of which the atomic NUCLEUS is composed, having significant mass but no electrical charge. (See also PROTON.)

NUCLEAR REACTOR A device which generates energy by FISSION in a controlled CHAIN REACTION.

NUCLEAR PILE — see PILE.

NUCLEAR WEAPON A weapon in which energy is released by FISSION in a runaway CHAIN REACTION. The term is also used to include a THER-

MONUCLEAR WEAPON, with the energy release being due to FUSION of nuclei.

NUCLEUS  The central part of an atom consisting of PROTONS and NEUTRONS. It has a positive electrical charge and contains most of the mass of the atom.

PERSHING MISSILE  A land-based medium-range single nuclear warhead missile of US manufacture.

PILE  A name originally given to nuclear reactors MODERATED by graphite blocks.

PLUTONIUM  A radioactive heavy metal element, not occurring naturally but produced in a NUCLEAR REACTOR by absorption of NEUTRONS in the non-fissile ISOTOPE U-238. Its radioactivity is due to the emission of ALPHA PARTICLES, which are harmful to humans but are of short range.

PLUTONIUM BOMB  A nuclear weapon based on the fission of PLUTONIUM leading rapidly to a runaway CHAIN REACTION. Such a bomb was dropped on Nagasaki in August 1945.

POLARIS  A submarine-launched intermediate-range ballistic missile of US manufacture.

PROTON  One of the fundamental particles which make up the atomic NUCLEUS. It has a positive electrical charge and a significant mass. (See also NEUTRON.)

PWR  A nuclear reactor MODERATED and cooled by LIGHT WATER, the latter being under pressure to prevent it boiling. The hot water circulates through a heat exchanger in which steam is generated to drive a turbine. Britain has one, known as Sizewell 'B', operating on the coast of Suffolk.

RADIOACTIVITY  Radioactivity is the natural or artificially induced disintegration of atomic NUCLEI which results in the emission of ionising rays (see ALPHA, BETA and GAMMA RAYS). This is often described as RADIATION, which can be harmful to living tissue and so some form of shielding is necessary to protect operating personnel.

RADIATION — see RADIOACTIVITY

REACTOR PRESSURE VESSEL  A steel or reinforced concrete container surrounding the CORE of a NUCLEAR REACTOR to maintain pressure on the COOLANT.

REFUELLING  Involves extracting the spent fuel elements first by withdrawing into shielded containers and removal to a COOLING POND. This is followed by inserting new fuel elements. This all takes time and the reactor normally has to remain shut down, although claims have been made for so-called 'on-line' refuelling with some reactor designs.

REPROCESSING  The extraction of PLUTONIUM from the various ISOTOPES present in SPENT FUEL. It requires a complex chemical processing plant with shielding to protect the operators from RADIATION.

SECONDARY NEUTRONS — see CHAIN REACTION

SKYBOLT    An air-to-ground nuclear missile, intended for the US Air Force, and which was to enable planes to launch it as much as 1600 kilometres from the target. Britain was to purchase Skybolt but the project was abandoned by the US government without advance notice.

SPENT FUEL    Nuclear fuel taken from the reactor at the end of its useful life.

SPENT FUEL STORAGE POND — see COOLING POND

STAND-OFF BOMB    A small rocket with a warhead, which may be nuclear, fired from an aircraft and designed to attack a target some distance ahead.

STEAM GENERATING HEAVY WATER REACTOR (SGHWR)    A reactor using HEAVY WATER as the MODERATOR, and which generates steam for power purposes. A 100 MW prototype was successfully operated by the UKAEA at Winfrith Heath in Dorset.

THERMAL REACTOR    A nuclear reactor in which NEUTRONS produced by FISSION are slowed down, or have their energy reduced, by collision with the NUCLEI of atoms in the MODERATOR.

THERMONUCLEAR WEAPON    One in which enormous quantities of energy are released by fusion of atomic nuclei at an extremely high temperature generated by an associated FISSION reaction.

THORP    A large REPROCESSING facility for separating PLUTONIUM from SPENT FUEL, operated at the Sellafield site by British Nuclear Fuels Limited.

TRIDENT    A submarine-launched BALLISTIC MISSILE, successor to POLARIS, only more powerful and technically advanced in direction and control.

TRITIUM    An ISOTOPE of hydrogen with a PROTON and two NEUTRONS in its NUCLEUS. The only isotope of hydrogen which is RADIOACTIVE. Used in the production of THERMONUCLEAR WEAPONS.

URANIUM    A heavy metallic element extracted from URANIUM ORE.

URANIUM BOMB    NUCLEAR WEAPON based on FISSION of the U-235 isotope. Was used over Hiroshima in 1945.

URANIUM ORE    Uranium is not obtained in the metallic state but there are ores of differing chemical composition, a common one being the oxide of formula $U^3O^8$. However, the oxide as mined is always present in the parent rock at a low concentration, rarely over 0.3 per cent. Has informally been referred to as 'Yellow Cake'.

VITRIFICATION    Processes aimed at incorporating HIGH LEVEL WASTE (possibly SPENT FUEL) in an insoluble, possibly glass-like material, for long-term underground storage at great depth.

# Appendix 2

# Short bibliography

Publications on nuclear energy are legion. Only a selected few are listed below, particularly some which made me think deeply about the topic without necessarily agreeing with the viewpoint taken by the authors.

Bacon, H. and Valentine, J., *Power Corrupts: The Arguments against Nuclear Power* (Pluto Press, London, 1981)

Blowers, A. and Pepper, D., (Eds.), *Nuclear Power in Crisis* (Croom Helm, London, 1987)

Breach, Ian, *Windscale Fallout* (Penguin, London 1978)

Burn, Duncan, *Nuclear Power and the Energy Crisis: Politics and the Atomic Industry* (Macmillan, London, 1978)

Clark, R.W., *The Birth of the Bomb* (The Scientific Book Club, London, 1961)

Cutler, J. and Edwards, R., *Britain's Nuclear Nightmare* (Sphere Books, London, 1988)

Davies, Philip, *Magnox: The Reckoning* (Friends of the Earth, London, 1980)

*Development of Atomic Energy 1939-1984: Chronology of Events* (UKAEA, London, 1984)

Foote, G., *A Chronology of Post War British Politics* (Croom Helm, London, 1988)

Freedman, Lawrence, *Britain and Nuclear Weapons* (Macmillan, London, 1980)

Gowing, Margaret, *Britain and Atomic Energy: 1939-45* (Macmillan, London, 1964)

Gowing, Margaret with Arnold, L., *Independence and Deterrence, Britain and Atomic Energy, 1945-52.* Vol. 1, *Policy Making*; Vol. 2, *Policy Execution* (UKAEA, Macmillan, London, 1974)

Jay, Kenneth, *Calder Hall* (Methuen, London, 1956)

Keegan, John, *Face of Battle* (final chapter) (Penguin, London, 1978)

*Nuclear Power and the Environment*, generally known as The Flowers Report; (6th Report of The Royal Commission on Environmental Pollution, Cmnd 6618, HMSO, 1976)

Patterson, Walter, C., *Nuclear Power* (Penguin, London, 1976)

Patterson, Walter C., *Going Critical: An Unofficial History of British Nuclear Power* (Grafton Books, London, 1985)

Pocock, R. F., *Nuclear Power: Its Development in the United Kingdom* (Unwin, London, 1977)

Prins, Gwyn, (Ed.) *Defended to Death; A Study of the Nuclear Arms Race*, (from the Cambridge University Disarmament Seminar) (Penguin Books, London, 1983)

*Review of Nuclear Power* (HM Government, May 1995)

*Report by Nuclear Installations Inspectorate on The Management of Safety at Windscale* (HMSO, February 1981)

Roberts, Alun, *The Rossing File: The Inside Story of Britain's Secret Contract for Namibian Uranium*, (CANUC, 1980)

Rotblat, Joseph, *Nuclear Radiation in Warfare* (SIPRI, Taylor and Francis, 1981)

Rotblat, Joseph, *Nuclear Reactors: To Breed or Not to Breed* (The Royal Society, London, 1977)

*The Church and the Bomb*, Working party report, Board for Social Responsibility of the Synod of Church of England; chair, Bishop of Salisbury (Hodder and Stoughton and CIO, 1982)

Thomas, Steve, *The Realities of Nuclear Power* (Cambridge University Press, 1988)

Valentine, John, *Atomic Crossroads: Before and after Sizewell* (Merlin Press, London, 1985)

Williams, Roger, *The Nuclear Power Decisions: British Politics 1953-78* (Croom Helm, 1980)

## Other useful sources of information:

### Periodicals

*Atom* (UKAEA monthly bulletin); *Nuclear Engineering International*, published monthly; *Bulletin of the Atomic Scientists*, published monthly.

### Year Books

Sipri Yearbook of World Armaments and Disarmament; CEGB Annual Reports and Statistical Yearbooks; UKAEA Annual Reports.

### Other material

Hansard; Biographies of PMs and Secretaries of State for Energy, Foreign Affairs, Defence, over the period covered by this book; The Benn Diaries.

# Index